Studies in Logic
Volume 5

Incompleteness in the Land of Sets

Volume 1
Proof Theoretical Coherence
Kosta Dosen and Zoran Petric

Volume 2
Model Based Reasoning in Science and Engineering
Lorenzo Magnani, editor

Volume 3
Foundations of the Formal Sciences IV: The History of the Concept of the Formal Sciences
Benedikt Löwe, Volker Peckhaus and Thoralf Räsch, editors

Volume 4
Algebra, Logic, Set Theory. Festschrift für Ulrich Felgner zum 65. Geburtstag
Benedikt Löwe, editor

Volume 5
Incompleteness in the Land of Sets
Melvin Fitting

Studies in Logic Series Editor
Dov Gabbay   dov.gabbay@kcl.ac.uk

Incompleteness in the Land of Sets

Melvin Fitting

© Individual author and King's College 2007. All rights reserved.

ISBN 978-1-904987-34-5
College Publications
Scientific Director: Dov Gabbay
Managing Director: Jane Spurr
Department of Computer Science
Strand, London WC2R 2LS, UK

Original cover design by Richard Fraser
Cover produced by orchid creative www.orchidcreative.co.uk
Printed by Lightning Source, Milton Keynes, UK

All rights reserved. No part of this publication may be reproduced, stored in a retrieval system or transmitted, in any form, or by any means, electronic, mechanical, photocopying, recording or otherwise, without prior permission, in writing, from the publisher.

PREFACE

Counting numbers are among the most concrete of mathematical objects. We all have certain intuitions concerning them—for instance, we all are pretty certain there is no biggest counting number. But intuitions can take us only so far, so rigorous methods have been developed over the years for studying the counting numbers (and other mathematical systems). We have *rules for calculation* with counting numbers, and we have *techniques of proof* concerning them. In the 1930's, Gödel, Turing, Post, Church, Tarski, Rosser, and others showed there were fundamental limitations on these techniques. This book is devoted to a presentation of their work. Essentially their results all combine to say there are profound restrictions on what can be known by our most reliable methods, applied to our most basic mathematical objects.

Each of the people just mentioned gave proofs of their theorems that are closely analogous to various well-known paradoxes. Tarski's proof is directly connected with the ancient paradox of the liar, which goes as follows. Consider the sentence S: "The sentence S is false." If S is true, what it asserts is correct, so S is false. Likewise if S is false, what S asserts is correct, so S is true. This is obviously impossible. What Tarski did was use, in a formally correct way, certain ideas underlying this paradox to establish limitations on what could be said about arithmetical truth in an arithmetical language. (A proper statement of his result will have to wait until the necessary groundwork has been laid.) Gödel noted that his own proof of what has come to be called the First Incompleteness Theorem was analogous to Richard's antinomy, which is a little more complicated. Let E be the collection of all decimals that can be defined using a finite number of words. E is countable, so its members can be listed in a sequence, say e_1, e_2, e_3,.... Now consider the number N whose decimal expansion is specified as follows: if the nth decimal place of e_n is p, let the nth decimal place of N be $p+1$ if $p \neq 9$, and 0 if $p = 9$. N can not be in E since N and e_n differ at decimal place n, for each n. Still, N has been defined using a finite number of words, so N must be in E. Gödel also pointed out that there was a close relationship between his argument and the liar paradox. Indeed, he made the interesting observation that, "Every epistemological antinomy can be used for a similar proof ...".

One of the best-known of the paradoxes invented in the twentieth century is that of Bertrand Russell. Call a set x *ordinary* if x does not belong to itself, that is, if $x \notin x$. Let S be the collection of all ordinary sets. Is S ordinary or not? If S is ordinary, it is in the collection of ordinary sets, so $S \in S$, which means S is not ordinary. Likewise, if S is not ordinary, it does not belong to the collection of ordinary sets, so $S \notin S$, which means S is ordinary. The discovery of this paradox eventually led to a precise formulation of set theory, and to the axiomatic systems in use today, but that is another story. In this book we have decided to use Russell's paradox to motivate in a uniform way all the incompleteness and undecidability results we prove. We begin by giving a brief hint of how this can be done.

Russell's paradox is about sets. Gödel and the other people mentioned above were primarily interested in counting numbers. Well, there is a fragment of set theory that is as concrete as arithmetic: the *hereditarily finite* sets. These consist of finite sets, of finite sets, of ... of finite sets. These are the sets that can be named using standard 'curly bracket' notation. We will see that there is an exact correspondence between counting numbers and hereditarily finite sets, so that in fact it does not matter which we choose to work with. Since the use of sets gives the appearance of greater flexibility, it is in this context that we carry out much of our work, though every result about the hereditarily finite sets immediately yields an analogous result about the counting numbers. If you know the terminology, our use of the hereditarily finite sets amounts to a form of Gödel numbering, introduced in full generality once and for all.

Once the necessary background has been developed concerning the hereditarily finite sets, the counting numbers, and the relationships between them, we move on to formal logic and set up a language L for talking about the hereditarily finite sets. Without going into the details now, a formula $\varphi(x)$ in the language L can be thought of as specifying a certain collection of hereditarily finite sets: the collection of sets s for which $\varphi(s)$ is true. In this way we can talk about sets and their definitions in L (shades of Richard's antinomy).

Next, it is well-known that all mathematics can be developed within the framework of set theory. In particular, finitary mathematical objects fit within the hereditarily finite portion of set theory. For example, it is possible to think of counting numbers as particular hereditarily finite sets. Indeed, even the elements of the language L can be thought of as hereditarily finite sets, and the formulas of L themselves can be taken to be certain special hereditarily finite sets. Then the formulas of L are *about* sets, and also *are* sets.

Now we can ask questions like: is the collection of formulas of L definable by a formula of L? (It turns out it is.) And we can reconstruct Russell's paradox as follows. Call a formula $\varphi(x)$ ordinary if it is not in the collection it defines. Let S be the collection of ordinary formulas. If S is definable, is the formula defining it ordinary or not? An argument exactly parallel to that for Russell's paradox above shows such a defining formula will be ordinary if and only if it is not. It follows that S, while defined informally in English, can not be defined by a formula of L. This is our first essential limitation result. From this it is a short step to Tarski's Theorem, and from that to a version of Gödel's Theorem.

In our discussion of Tarski's Theorem we made use of the notion of truth: $\varphi(x)$ defines the collection of sets s for which $\varphi(s)$ is *true*. Suppose we replace this use of truth by a notion of *provability* in a formal system. We will see that the argument based on Russell's paradox quite natuarlly divides in two, and one version leads to Church's Theorem while the other leads to Gödel's First Incompleteness Theorem more-or-less as Gödel conceived it. Further modifications to the same argument yield Post's and Rosser's Theorems. Internalizing Gödel's argument leads to his Second Incompleteness Theorem, and then to Löb's Theorem—our approach to these is through the convenient machinery of modal logics which, as it happens, were first brought into this area by Gödel himself. It is with this that the book ends.

While sets are fundamental, arithmetic is not neglected. Sets themselves are Gödel numbered, and this makes it easy to transfer results between arithmetic and set theory. This is not a book on recursion theory, but an attempt is made to establish the fundamental nature of the concepts that recursion theory treats. Here being Σ is taken as basic—recursive enumerability is defined as being Σ in an arithmetic model. The importance of being Σ is partly proved and partly discussed. We prove that whether or not a relation is Σ does not depend on whether we work with numbers or sets, and if with sets it does not depend on whether we work with them directly or via their Gödel numbers. Informally we discuss connections with computer programs and with intuitive notions of effectivity.

It must be emphasized that ours is an inherently mathematical approach. One designed for computer scientists would be better off taking words over a finite alphabet as basic. Of course finally all these approaches are equivalent, but emphasis differs, and for good reasons. Choices must be made, and in this book mathematics won.

The work presented here presupposes prior knowledge of basic first-order logic, including models, formal proofs, and the Gödel Completeness Theorem. Beyond this, mathematical sophistication is the chief need.

CONTENTS

PREFACE	v
CHAPTER 1 ON NUMBERS AND SETS	**1**
1 Introduction	1
2 Numbers	1
3 Set Theory Background	2
4 Hereditarily Finite Sets	5
5 Numbers and Sets	8
CHAPTER 2 LOGIC BACKGROUND	**11**
1 Introduction	11
2 Syntax	11
3 Models	14
4 Truth in Canonical Models	18
5 Representable Sets and Relations	20
CHAPTER 3 REPRESENTABILITY FOR SET THEORY	**23**
1 Introduction	23
2 When is a formula simple?	23
3 Formulas as program modules	26
4 Doing Away With Terms	27
5 Some More Δ_0 Sets	28
6 A Normal Form Theorem	31
7 Numbers are Δ_0	33
8 Finite Sequences and Arithmetic	35
CHAPTER 4 REPRESENTABILITY FOR ARITHMETIC	**39**
1 Basics	39
2 Arithmetic in Set Theory	40
3 Gödel's β function	42
4 Set Theory in Arithmetic	45
5 Σ is Σ	46

CHAPTER 5 TARSKI'S THEOREM OR REPRESENTING REPRESENTABILITY — 49
1. What is a Symbol? — 49
2. Concatenation — 50
3. Representing Terms — 52
4. Representing Formulas — 53
5. Substitution — 54
6. Representing Representability — 55
7. Russell's Paradox — 57
8. Tarski's Theorem — 58
9. Liars and Fixed Points — 59
10. Tarski's Theorem, Continued — 61

CHAPTER 6 COMPUTABILITY — 65
1. The Importance of Being Σ — 65
2. A Σ set, but don't ask Π — 69
3. Σ truth is Σ — 71
4. Kleene's Normal Form Theorem — 73
5. A Turing Machine Sketch — 75

CHAPTER 7 AXIOMATICS — 81
1. Introduction — 81
2. Truth in Models — 81
3. Formal Proofs — 83
4. Equality — 85
5. Theories — 87
6. Examples of Formal Theories — 91
6.1. Formal Arithmetic Theories — 91
6.2. Formal Set Theories — 92
6.3. Theories You Might Not Have Expected — 94

CHAPTER 8 GÖDEL'S THEOREM — 97
1. Gödel's Theorem, Tarski's Proof — 97
2. Finitary Proofs — 98
3. Representability in a Theory — 99
4. Ordinary Splits — 102
5. Gödel's Theorem, Gödel's Proof (more or less) — 103
6. ω-Consistency — 108

CHAPTER 9   CHURCH'S THEOREM, ROSSER'S THEOREM   111
1 Introduction 111
2 Church's Theorem 111
3 Rosser's Theorem 114
4 Rosser's Theorem Continued 116

CHAPTER 10   GÖDEL'S SECOND THEOREM   123
1 Introduction 123
2 The Gödel Fixed Point Theorem 123
3 The Löb Provability Conditions 126
4 Abbreviated Notation 127
5 Gödel's Second Incompleteness Theorem 129
6 Löb's Theorem 130
7 Gödel-Löb logic, GL 132

FURTHER READING   135

INDEXES   138

CHAPTER 1

ON NUMBERS AND SETS

1 Introduction

This chapter is a quick and informal survey of the properties of counting numbers and hereditarily finite sets. We begin with numbers, a part which is short since everybody believes they know about them already. We next discuss general notions from set theory, then narrow things to the hereditarily finite sets themselves. Finally we discuss some connections between sets and numbers. The formal work starts with the next chapter.

2 Numbers

Since the only numbers that we will use are the non-negative integers, we just call them numbers, to keep terminology simple.

Definition 2.1. By *numbers* we mean members of $\{0, 1, 2, \ldots\}$.

For us the basic operations on numbers are: counting (or successor), addition, multiplication, and sometimes exponentiation. The basic relations on numbers are equality and less-than, and by extension less-than-or-equal, greater-than, etc. We assume everybody is familiar with these notions.

The basic insight concerning numbers is the induction principle. This is our chief tool for establishing things about all members of the infinite set of numbers.

Definition 2.2. A set of numbers is *inductive* if it contains 0 and is closed under successor.

Induction Principle There is exactly one inductive set, the set of numbers itself.

We give an example of the use of this principle. Suppose we are told there is a function f from numbers to numbers such that $f(0) = 1$ and $f(n+1) = 2 \times f(n)$. We would like to show that, in fact, $f(n) = 2^n$ for all numbers n.

Let S be the set of numbers n for which $f(n) = 2^n$; we show S is inductive. First of all, $f(0) = 1$ and $2^0 = 1$ so $0 \in S$. Next, suppose $k \in S$, so that

$f(k) = 2^k$. Then $f(k+1) = 2 \times f(k) = 2 \times 2^k = 2^{k+1}$, so $k+1 \in S$. Thus S is inductive, so S is the set of numbers, and so $f(n) = 2^n$ for all numbers n.

Induction proofs are used constantly here. We assume you are generally familiar with them.

Exercises

Exercise 2.1. Call a set T of numbers *completely inductive* if, for any number n, $n \in T$ provided all numbers smaller than n are in T. Use the Induction Principle and show there is exactly one completely inductive set, the set of numbers itself.

3 Set Theory Background

Now we turn our attention to set theory, beginning with general principles. There are standard devices for treating numbers, relations, functions, and other mathematical objects as sets. This section is a brief review of such matters. As we remarked earlier, our discussion is an informal one. We assume that if $x_1, x_2, \ldots, x_n$ are sets, so is $\{x_1, x_2, \ldots, x_n\}$. We assume we can form unions, and that the empty set exists. Likewise we assume $\in$ and $\subseteq$ are generally meaningful. Notation is standard.

Since the treatment of both functions and relations will involve the notion of ordered pair, we begin here. The crucial fact about ordered pairs is:

$$\langle a, b \rangle = \langle x, y \rangle \quad \text{iff} \quad a = x \text{ and } b = y \qquad (1.1)$$

The following device (due to Kuratowski) designates a certain set to act as an ordered pair, in such a way that (1.1) is demonstrably true.

Definition 3.1.
$$\langle a, b \rangle = \{\{a\}, \{a, b\}\}$$

Theorem 3.2. $\langle a, b \rangle = \langle x, y \rangle$ *if and only if* $a = x$ *and* $b = y$.

Proof If $a = x$ and $b = y$, trivially $\langle a, b \rangle = \langle x, y \rangle$. Now suppose $\langle a, b \rangle = \langle x, y \rangle$, that is,

$$\{\{a,\}, \{a, b\}\} = \{\{x\}, \{x, y\}\}.$$

We refer to the left-hand and the right-hand sides in this equation as LHS and RHS respectively. The proof splits into three cases.

Case 1. Suppose $a = b$. Then LHS $= \{\{a\}, \{a, a\}\} = \{\{a\}\}$. This has only one member, hence RHS also has one member. Then $\{x\} = \{x, y\}$, so $x = y$. Now RHS $= \{\{x\}, \{x, x\}\} = \{\{x\}\}$. We have $\{\{a\}\} = \{\{x\}\}$, so $\{a\} = \{x\}$, so $a = x$. In this case $a = b = x = y$.

Case 2. Suppose $x = y$. This is similar to Case 1.

Case 3. Suppose $a \neq b$ and $x \neq y$. Now, $\{a\} \in$ LHS so $\{a\} \in$ RHS. This implies $\{a\} = \{x\}$ or $\{a\} = \{x, y\}$. The second of these is impossible since $\{a\}$ has one member but $\{x, y\}$ has two. Consequently $\{a\} = \{x\}$, so $a = x$. Further, $\{a, b\} \in$ LHS, so $\{a, b\} \in$ RHS. Then $\{a, b\} = \{x\}$ or $\{a, b\} = \{x, y\}$. This time the first is impossible, since $a \neq b$, so the second holds. Then, since $b \in \{a, b\}$ we have $b \in \{x, y\}$. Either $b = x$ or $b = y$. The first is impossible, since $a = x$ but $a \neq b$. So $b = y$ and we are done. ∎

We needn't go on to discuss ordered triples, quadruples, etc. in such a detailed fashion, since these can be reduced to the notion above.

Definition 3.3. We make the following sequence of definitions.

$$\langle x, y, z \rangle = \langle x, \langle y, z \rangle \rangle$$

$$\langle x, y, z, w \rangle = \langle x, \langle y, z, w \rangle \rangle$$

and so on.

Next we discuss relations. To each intuitively conceived relation R there corresponds a set of ordered pairs R^*: put $\langle x, y \rangle$ in R^* if x is in the relation R to y. Likewise to each set R^* of ordered pairs there naturally corresponds a relation R: let x be in the relation R to y just when $\langle x, y \rangle \in R^*$. Consequently, in set theory one thinks of relations as *being* sets of ordered pairs.

Definition 3.4. A *binary relation* is a set R of ordered pairs. A *three-place relation* is a set of ordered triples, and so on.

Functions too may be treated as sets of ordered pairs, meeting a special condition. Suppose f is an intuitively conceived function. Define a set f^* by: put $\langle x, y \rangle$ in f^* if $f(x) = y$. Then f^* has the *single-valuedness* property: if $\langle x, y \rangle \in f^*$ and also $\langle x, y' \rangle \in f^*$ then $y = y'$. But this can be reversed. Suppose f^* is a set of ordered pairs (that is, a relation) meeting the single-valuedness condition. We may then define a function f by setting $f(x)$ to be that unique y such that $\langle x, y \rangle \in f^*$. This leads to the following.

Definition 3.5. An *n-place function* is an $n + 1$-place relation f that meets the single-valuedness condition:

$$\langle x_1, \ldots, x_n, y \rangle \in f \text{ and } \langle x_1, \ldots, x_n, y' \rangle \in f \text{ implies } y = y'.$$

If f is an n-place function we write $f(x_1, \ldots, x_n) = y$ for $\langle x_1, \ldots, x_n, y \rangle \in f$. The *domain* of f is the set of all n-tuples $\langle x_1, \ldots, x_n \rangle$ such that, for some

y, $\langle x_1, \ldots, x_n, y \rangle \in f$. Likewise the *range* of f is the set of all y such that for some $\langle x_1, \ldots, x_n \rangle$, $\langle x_1, \ldots, x_n, y \rangle \in f$.

Numbers can be represented as sets in many ways. It is particularly convenient to represent the number n by some set having n members, and Von Neumann introduced a simple way of doing so. Suppose that distinct sets have been assigned to each of the numbers $0, 1, 2, \ldots, n-1$, say these representing sets are $0^*, 1^*, 2^*, \ldots, (n-1)^*$. We need a handy n-element set to represent n; but there is an obvious one already present: let n^* be $\{0^*, 1^*, 2^*, \ldots, (n-1)^*\}$. That is, the representative for a number is simply taken to be the set of representatives of all smaller numbers. Von Neumann's idea is to make this the very definition of number,' in a set theoretic context. But then 0 would be $\emptyset$, and $n+1 = \{0, 1, \ldots, n-1, n\} = \{0, 1, \ldots, n-1\} \cup \{n\} = n \cup \{n\}$. So our definitions are as follows.

Definition 3.6. An operation on sets is defined by: $x^+ = x \cup \{x\}$. Then we set $0 = \emptyset$, $1 = 0^+$, $2 = 1^+$, $3 = 2^+$, and so on. Finally, $\omega = \{0, 1, 2, \ldots\}$. Equivalently, ω is the smallest set that contains $\emptyset$ and is closed under $x \mapsto x^+$.

It follows that for members of ω, the operation $<$ corresponds to $\in$. That is, if $n < k$ as numbers, then as members of ω, $n \in k$, and conversely. This yields the following important principle.

Trichotomy Principle For $n, k \in \omega$, exactly one of: $n \in k$, $n = k$, or $k \in n$.

Sometimes we take numbers as primitive. Sometimes we think of them as sets—members of ω—in the manner given above. The uses can be distinguished by context (we hope).

Finite sequences can be treated quite easily now. Instead of the 'conventional' $a_0, a_1, \ldots, a_n$, we use the function f, whose domain is $\{0, 1, \ldots, n\}$, such that $f(i) = a_i$. But $\{0, 1, \ldots, n\}$ is just n^+, a number, and functions are particular sets.

Definition 3.7. By a *finite sequence* we mean any function f whose domain is a number, called the *length* of the finite sequence. If f is a finite sequence, we generally write f_i in place of $f(i)$.

These are all the notions from general set theory we will need for now.

Exercises

Exercise 3.1. Prove $\langle a, b, c \rangle = \langle x, y, z \rangle$ if and only if $a = x$, $b = y$ and $c = z$.

Exercise 3.2. Suppose a and b are two finite sequences, each of length 2, and $a_0 = b_0$ and $a_1 = b_1$. Prove $a = b$. Note: this is equality between sets, so what must be proved is that a and b have the same members.

4 Hereditarily Finite Sets

Like numbers, the hereditarily finite sets are among the most concrete of mathematical objects. An hereditarily finite set is a set which is finite, all of whose members are finite, with their members finite, and so on. When mathematical entities like relations and functions are represented by sets, it is the ones we think of as finitary that get represented as hereditarily finite sets. In particular, all the syntactic structures of formal logic, like formulas and proofs, can be naturally represented as hereditarily finite sets, and this is the source of our interest in them here. In this section we give two different characterizations of the hereditarily finite sets, and demonstrate their equivalence.

Our first characterization is closest to the informal notion of a finite set of finite set of finite sets of, etc. In fact it is that definition backwards, telling how hereditarily finite sets may be built up, rather than broken down. We use the name H for now; later it will be replaced by a better one.

Definition 4.1. The collection H of *hereditarily finite sets* is the collection generated by the following rules.

1. $\emptyset \in H$,
2. if $x_1, x_2, \ldots, x_n \in H$ then $\{x_1, x_2, \ldots, x_n\} \in H$.

Notice that every finite subset of H is a member of H. This is true for the empty subset by 1, and for non-empty subsets by 2.

Our second characterization will involve the power set operation: $\mathcal{P}(x)$ denotes the collection of all subsets of x.

Definition 4.2. We define a sequence of sets, $R_0, R_1, \ldots$, and a 'limit' term as follows.

1. $R_0 = \emptyset$,
2. $R_{n+1} = \mathcal{P}(R_n)$,
3. $R_\omega = R_0 \cup R_1 \cup R_2 \cup \ldots$.

We have characterized two interesting collections, H, and R_ω. We now show them to be the same thing.

Theorem 4.3. *For each n, $R_n \subseteq H$, and consequently $R_\omega \subseteq H$.*

Proof We show by induction that for each n, $R_n \subseteq H$. That $R_\omega \subseteq H$ then follows by the definition of R_ω.

$R_0 \subseteq H$ since $R_0 = \emptyset$.

Suppose $R_n \subseteq H$; we show $R_{n+1} \subseteq H$. Let $x \in R_{n+1}$. Then $x \subseteq R_n$, so $x \subseteq H$. But R_n is finite, by Exercise 4.2, so x is finite. Since it is a finite subset of H, it must be a member of H. ∎

Exercise 4.3 shows that the sequence $R_0, R_1, \ldots$ is cumulative, strictly increasing, and eventually includes each number. The following shows it eventually includes *every* hereditarily finite set.

Theorem 4.4. $H \subseteq R_\omega$.

Proof Our definition of the hereditarily finite sets really is a list of instructions for generating them. Now we show that when we follow those instructions the sets we produce all belong to R_ω.

First, $\emptyset \in H$. But also, $\emptyset \subseteq R_0$, so $\emptyset \in R_1$, and hence $\emptyset \in R_\omega$.

Next, suppose we have generated hereditarily finite sets $x_1, x_2, \ldots, x_k$ and, as we did so we verified that each one is in R_ω. The definition of H says $\{x_1, x_2, \ldots, x_k\} \in H$. We show it is also in R_ω. Well, $x_1 \in R_\omega$, so for some number n_1, $x_1 \in R_{n_1}$. Likewise $x_2 \in R_{n_2}$ for some n_2, ..., $x_k \in R_{n_k}$ for some n_k. Let n be the largest of $n_1, n_2, \ldots, n_k$. Since the R_i sequence is cumulative, by Exercise 4.3, we have $\{x_1, x_2, \ldots, x_k\} \subseteq R_n$, so $\{x_1, x_2, \ldots, x_k\} \in R_{n+1}$, and thus $\{x_1, x_2, \ldots, x_k\} \in R_\omega$. ∎

We now know that R_ω is exactly the collection H of hereditarily finite sets. From now on we abandon the name H, and use R_ω exclusively to designate the collection of hereditarily finite sets. Pleasantly, the characterization of R_ω by its 'stages of construction' gives us certain useful additional information. For example, it allows us to introduce the notion of *rank*. Since the R_i sequence is cumulative, if $x \in R_\omega$, there must be a *first* R_n to which x belongs.

Definition 4.5. An hereditarily finite set x has *rank* n if $x \in R_{n+1}$ but $x \notin R_n$. That is, R_{n+1} is the first R_i to which x belongs.

Note that Exercise 4.3 says that if n is a number, its rank is n.

Suppose x is of rank k, so R_{k+1} is the first R_i to which x belongs. Since $x \in R_{k+1}$, $x \subseteq R_k$. Now, if $y \in x$, $y \in R_k$, so the first R_i to which y belongs must be R_k or an earlier term. We have established the following.

Useful Fact If x is hereditarily finite, the rank of any member of x must be less than the rank of x.

Chapter 1. On Numbers and Sets

This will come in handy for proving things about members of R_ω. We conclude this section with a list of properties of R_ω. These either follow from exercises below, or are easy consequences of the definition directly.

1. $x \in R_\omega$, $y \in x$ implies $y \in R_\omega$. That is, a member of an hereditarily finite set is hereditarily finite; R_ω is transitive.

2. $x \in R_\omega$, $y \subseteq x$ implies $y \in R_\omega$. That is, a subset of an hereditarily finite set is hereditarily finite.

3. $\emptyset \in R_\omega$.

4. R_ω is closed under the formation of finite subset, power set, and union.

5. If $x \in R_\omega$, x is disjoint from one of its members.

Exercises

Exercise 4.1. Show the following twice, once using Definition 4.1 and once using Definition 4.2.

1. The number 3 is in R_ω.
2. If $x \in R_\omega$ then $x^+ \in R_\omega$.
3. Every number is in R_ω.

Exercise 4.2. Show the following, to justify the name 'power set'.

1. Suppose every n-element set has 2^n subsets. Let x have $n+1$ elements. Prove $\mathcal{P}(x)$ has 2^{n+1} elements.
2. Prove that an n-element set has 2^n subsets.
3. Prove that R_n is finite, for each number n.

Exercise 4.3. Prove each of the following.

1. If $x \in R_n$ then $x \subseteq R_n$ (use induction on n).
2. $R_n \in R_{n+1}$.
3. $R_n \subseteq R_{n+1}$.
4. $n \in R_{n+1}$ (use induction on n).
5. $n \notin R_n$ (again, use induction).

Exercise 4.4. Prove the following is true of R_ω: any member x has a member y disjoint from it, i.e. such that $x \cap y = \emptyset$. Hint: take y to be a member of x of lowest rank.

Exercise 4.5. Prove each of the following:

1. $x, y \in R_n$ implies $\{x, y\} \in R_{n+1}$.
2. $x \in R_n$ implies $\mathcal{P}(x) \in R_{n+1}$.
3. $x \in R_n$ implies $\bigcup x \in R_n$.

5 Numbers and Sets

We have seen in Section 3 that numbers can be identified with particular sets, and in Section 4 that these sets are hereditarily finite. Now we make this identification official policy.

- From now on, numbers are taken to be the members of ω.

In effect, when we investigate the hereditarily finite sets, we are investigating numbers, since $\omega \subseteq R_\omega$. This is the first, but not the last, instance in this book of an identification of some intuitive notion with a family of sets.

In fact, the relationship between the hereditarily finite sets and numbers is stronger than the preceding would suggest. There is a useful *isomorphism* between them, due to Ackermann. That is, hereditarily finite sets can be uniquely coded by numbers in such a way that basic manipulations on sets correspond to easily characterizable arithmetic manipulations. This isomorphism will play a fundamental role in much that follows. For simplicity of presentation, we begin with the inverse of the coding. The idea is very elegant, and makes use of binary notation. Suppose we number the positions of digits that appear in a base 2 name, from right to left, starting with 0. For example, writing position numbers as subscripts, 10010 has positions numbered as follows: $1_4 0_3 0_2 1_1 0_0$

Definition 5.1. A function $\mathcal{H} : \omega \to R_\omega$ is specified by the following.

$\mathcal{H}(n) = \{\mathcal{H}(k) \mid 1 \text{ occurs in position } k \text{ in the base 2 expansion of } n\}$.

Using this schema, the first few items are as follows.

$$\begin{aligned}
\mathcal{H}(0) &= \emptyset \\
\mathcal{H}(1) &= (1)_2 = \{\mathcal{H}(0)\} = \{\emptyset\} \\
\mathcal{H}(2) &= (10)_2 = \{\mathcal{H}(1)\} = \{\{\emptyset\}\} \\
\mathcal{H}(3) &= (11)_2 = \{\mathcal{H}(0), \mathcal{H}(1)\} = \{\emptyset, \{\emptyset\}\}
\end{aligned}$$

The following is central, and not difficult. Its proof is left as an exercise.

Theorem 5.2. *The mapping $\mathcal{H} : \omega \to R_\omega$ is one-to-one and onto.*

Definition 5.3. The mapping $\mathcal{G} : R_\omega \to \omega$ is the inverse of $\mathcal{H}$.

The mapping $\mathcal{G}$ codes hereditarily finite sets by numbers. We will call $\mathcal{G}(s)$ the *Gödel number* of the set s. Now we show that set-theoretic calculations correspond easily to number-theoretic ones on their Gödel numbers. In the following, DIV means quotient of a division, MOD is the remainder.

Proposition 5.4. *For $s, t \in R_\omega$,*
$s \in t$ if and only if $[\mathcal{G}(t) \text{ DIV } 2^{\mathcal{G}(s)}] \text{ MOD } 2 = 1$.

Proof $(n \text{ DIV } 2^k) \text{ MOD } 2$ is the digit in position k of the binary expansion of n. ∎

In the next item, BITAND is the operation that combines two binary expansions by 'and'ing them, place by place. More precisely, n BITAND m is the number whose kth binary digit is 1 if the kth binary digits of both n and m are 1, and otherwise is 0. In a similar way, BITOR is the operation that 'or's binary expansions. Now, the following should be straightforward.

Proposition 5.5. *For $s, t \in R_\omega$:*

1. $\mathcal{G}(s \cap t) = \mathcal{G}(s) \text{ BITAND } \mathcal{G}(t)$.
2. $\mathcal{G}(s \cup t) = \mathcal{G}(s) \text{ BITOR } \mathcal{G}(t)$.
3. $\mathcal{G}(\{t\}) = 2^{\mathcal{G}(t)}$.
4. $\mathcal{G}(s \cup \{t\}) = \mathcal{G}(x) \text{ BITOR } 2^{\mathcal{G}(t)}$.

The operations of BITAND and BITOR can, in turn, be reduced to more conventional ones, plus recursion. We leave this to you in the exercises.

Exercises

Exercise 5.1. Taking 3 to be a member of R_ω, evaluate $\mathcal{G}(3)$.

Exercise 5.2. Prove the mapping $\mathcal{H} : \omega \to R_\omega$ is one-to-one and onto. Hint: to show the mapping is onto, show every member of R_n is in the range, by induction on n. For this the Useful Fact from the previous section will be useful. Likewise, if $\mathcal{H}$ is not one-to-one, there must be a hereditarily finite set of lowest rank that is the image of two numbers, Now derive a contradiction.

Exercise 5.3. Prove the following.

1. n BITAND $m = ((n$ DIV $2)$ BITAND $(m$ DIV $2)) \times 2 + (n$ MOD $2) \times (m$ MOD $2)$.

2. n BITOR $m = ((n$ DIV $2)$ BITOR $(m$ DIV $2)) \times 2 + (n$ MOD $2) + (m$ MOD $2) - (n$ MOD $2) \times (m$ MOD $2)$.

Exercise 5.4. Prove that $\mathcal{G}(x^+) = 2^{\mathcal{G}(x)} + x$.

Exercise 5.5. Show that, if $s \in t$ then $\mathcal{G}(s) < \mathcal{G}(t)$.

CHAPTER 2

LOGIC BACKGROUND

1 Introduction

Perhaps the main topic of this book is the distinction between what is true and what can be formally proved. Of course before something can be proved, it must be said. In this chapter we set up the syntax of first-order logic, to make this precise. Then we quickly discuss semantics, so that a rigorous notion of truth is available. Provability will be discussed starting in Chapter 7. Even if you are already familiar with the material of this chapter, skim through it quickly to make sure you get the notation straight. Books differ on this.

2 Syntax

We will be using a couple of first-order languages here. Certain symbols are common to all of them; we begin with these.

Propositional Connectives The choice here is somewhat arbitrary. We use $\neg$ (not) and the *binary connectives* $\wedge$ (and), $\vee$ (or), $\supset$ (implies) and sometimes $\equiv$ (if and only if).

Quantifiers There are two: $\forall$ (for all, the universal quantifier), and $\exists$ (there exists, the existential quantifier).

Punctuation ')', '(', and ','.

Variables $v_0, v_1, \ldots$ (informally we will generally write $x, y, \ldots$, for reading ease).

Now we come to the symbols that can vary from language to language.

Definition 2.1. A *first order language* is determined by specifying:

1. a finite or countable set **R** of *relation symbols*, each of which has a positive integer associated with it. If $P \in \mathbf{R}$ has the integer n associated with it, we say P is an n-place relation symbol;

2. a finite or countable set **F** of *function symbols*, each of which has a positive integer associated with it. If $f \in \mathbf{F}$ has the integer n associated with it, we say f is an n-place function symbol;

3. a finite or countable set **C** of *constant symbols*.

We use the notation $L(\mathbf{R}, \mathbf{F}, \mathbf{C})$ for the first-order language determined by **R**, **F** and **C**.

Having specified the basic element of syntax, the alphabet, we go on to more complex constructions. We begin with *terms* which are, in a sense, the nouns and pronouns of the language. They are the things that name objects, at least intuitively.

Definition 2.2. The family of *terms* of $L(\mathbf{R}, \mathbf{F}, \mathbf{C})$ is the smallest set meeting the conditions:

1. any variable is a term of $L(\mathbf{R}, \mathbf{F}, \mathbf{C})$;

2. any constant symbol (member of **C**) is a term of $L(\mathbf{R}, \mathbf{F}, \mathbf{C})$;

3. if f is an n-place function symbol (member of **F**) and $t_1, \ldots, t_n$ are terms of $L(\mathbf{R}, \mathbf{F}, \mathbf{C})$, then $f(t_1, \ldots, t_n)$ is a term of $L(\mathbf{R}, \mathbf{F}, \mathbf{C})$.

A term is *closed* if it contains no variables.

Next we turn to *formulas*, the things that are supposed to make assertions.

Definition 2.3. An *atomic formula* of $L(\mathbf{R}, \mathbf{F}, \mathbf{C})$ is any expression of the form $R(t_1, \ldots, t_n)$ where R is an n-place relation symbol (member of **R**) and $t_1, \ldots, t_n$ are terms of $L(\mathbf{R}, \mathbf{F}, \mathbf{C})$.

Definition 2.4. The family of *formulas* of $L(\mathbf{R}, \mathbf{F}, \mathbf{C})$ is the smallest set meeting the conditions:

1. any atomic formula of $L(\mathbf{R}, \mathbf{F}, \mathbf{C})$ is a formula of $L(\mathbf{R}, \mathbf{F}, \mathbf{C})$;

2. if A is a formula of $L(\mathbf{R}, \mathbf{F}, \mathbf{C})$ so is $\neg A$;

3. for a binary connective $\circ$, if A and B are formulas of $L(\mathbf{R}, \mathbf{F}, \mathbf{C})$, so is $(A \circ B)$;

4. if A is a formula of $L(\mathbf{R}, \mathbf{F}, \mathbf{C})$ and x is a variable, then $(\forall x)A$ and $(\exists x)A$ are formulas of $L(\mathbf{R}, \mathbf{F}, \mathbf{C})$.

It is easy to say which terms are to be thought of as naming objects: the closed terms, without variables. In a similar way, formulas without "real" variables are the ones we will think of as making assertions. The trouble is, variables don't count if a quantifier governs them. We need to specify which variable occurrences are *bound*, that is, subject to a quantifier, and which are *free*, that is, not subject to a quantifier. There are a few, slightly different, ways this can be done. We find it convenient to first define the notion of *substitution*, since we will need this anyway. The notation we use is $\begin{bmatrix} x \\ t \end{bmatrix}$, to denote the replacement of the variable x by the term t. We begin with substitution in terms, where things are pretty straightforward anyway. The definition parallels that of term itself.

Definition 2.5. Let x be a variable and t be a term of $L(\mathbf{R}, \mathbf{F}, \mathbf{C})$:

1. for a variable y,
$$y \begin{bmatrix} x \\ t \end{bmatrix} = \begin{cases} t & \text{if } x = y \\ y & \text{if } x \neq y; \end{cases}$$

2. for a constant symbol c, $c \begin{bmatrix} x \\ t \end{bmatrix} = c$;

3. for a function symbol f, $f(t_1, \ldots, t_n) \begin{bmatrix} x \\ t \end{bmatrix} = f(t_1 \begin{bmatrix} x \\ t \end{bmatrix}, \ldots, t_n \begin{bmatrix} x \\ t \end{bmatrix})$.

It is straightforward to show that for terms, $u \begin{bmatrix} x \\ t \end{bmatrix}$ is simply the result of replacing *all* occurrences of x in u with occurrences of t (Exercise 2.1). For formulas, things are made more complicated by the presence of quantifiers.

Definition 2.6. Substitution in formulas of $L(\mathbf{R}, \mathbf{F}, \mathbf{C})$ is characterized as follows.

1. For an atomic formula, $R(t_1, \ldots, t_n) \begin{bmatrix} x \\ t \end{bmatrix} = R(t_1 \begin{bmatrix} x \\ t \end{bmatrix}, \ldots, t_n \begin{bmatrix} x \\ t \end{bmatrix})$.

2. $[\neg A] \begin{bmatrix} x \\ t \end{bmatrix} = \neg A \begin{bmatrix} x \\ t \end{bmatrix}$.

3. For a binary connective $\circ$, $(A \circ B) \begin{bmatrix} x \\ t \end{bmatrix} = (A \begin{bmatrix} x \\ t \end{bmatrix} \circ B \begin{bmatrix} x \\ t \end{bmatrix})$.

4. For a quantified formula,
$$[(\forall y) A] \begin{bmatrix} x \\ t \end{bmatrix} = \begin{cases} (\forall y) A & \text{if } x = y \\ (\forall y)[A \begin{bmatrix} x \\ t \end{bmatrix}] & \text{if } x \neq y. \end{cases}$$

Similarly for the existential quantifier.

Now we can say that the 'real' variables are the ones that change under substitution.

Definition 2.7. The variable x has a *free occurrence* in the formula A if $A \begin{bmatrix} x \\ t \end{bmatrix} \neq A$ for some term t. A formula in which no variables have free occurrences is called *closed*, or is said to be a *sentence*.

Exercise 2.3 asks for an alternative approach to the business of defining closed formulas. It will be convenient in Section 4 of Chapter 5 to have it available.

Notation Convention To keep notation simple, we will often indicate substitutions informally in the following way. If, at some point, we write $\varphi(x)$, it is assumed we are talking about a formula φ in which the variable x may have free occurrences (though it is not necessary that it have them). Then if, at some later point, we write $\varphi(t)$, this should be taken to designate the formula $\varphi(x)\left[\begin{smallmatrix}x\\t\end{smallmatrix}\right]$.

Exercises

Exercise 2.1. Show by induction on complexity that, for terms, $u\left[\begin{smallmatrix}x\\t\end{smallmatrix}\right]$ is the result of replacing all occurrences of x in u with occurrences of t.

Exercise 2.2. Suppose that only the variable x has free occurrences in the formula $\varphi(x)$, and that t is a closed term. Show that $\varphi(t)$ is a closed formula.

Exercise 2.3.

1. Without using the notion of substitution, give a (recursive) characterization of the relation: S is the set of variables that have free occurrences in the formula A.

2. Prove your characterization is equivalent to that of Definition 2.7.

3. Give a definition of sentence that does not use substitution.

3 Models

A formal language, as defined in the previous section, includes constant, function, and relation symbols. But expressions in such a language have no meaning until it is said what things these symbols designate. A *model* or *structure* is just such a designation. We briefly present the general notion of structure, then we introduce the two we are primarily interested in: a structure for arithmetic, and a structure for the hereditarily finite sets.

Definition 3.1. Let $L(\mathbf{R}, \mathbf{F}, \mathbf{C})$ be a first-order language. A *model* for this language is a structure $\mathcal{M} = \langle \mathcal{D}, \mathcal{I} \rangle$, where $\mathcal{D}$ is a non-empty set, called the *domain* of the model, and $\mathcal{I}$ is a mapping, called the *interpretation*, that: assigns to each n-place relation symbol $R \in \mathbf{R}$ some n-place relation $R^{\mathcal{I}}$ on $\mathcal{D}$; assigns to each n-place function symbol $f \in \mathbf{F}$ some n-place function $f^{\mathcal{I}} : \mathcal{D}^n \to \mathcal{D}$; and assigns to each constant symbol $c \in \mathbf{C}$ some member $c^{\mathcal{I}}$ of $\mathcal{D}$.

In short, a model specifies what objects we are talking about and what the relation, function and constant symbols mean when talking about these objects. Models specify meanings for basic symbols, but by an easy extension they can be made to specify meanings for closed terms as well.

Definition 3.2. Let $L(\mathbf{R}, \mathbf{F}, \mathbf{C})$ be a language and $\mathcal{M} = \langle \mathcal{D}, \mathcal{I} \rangle$ be a model for the language. To each closed term t is assigned a 'meaning' $t^{\mathcal{M}} \in \mathcal{D}$ as follows.

1. For a constant symbol c, set $c^{\mathcal{M}} = c^{\mathcal{I}}$.

2. For an n-place function symbol f, and closed terms $t_1, \ldots, t_n$, set $[f(t_1, \ldots, t_n)]^{\mathcal{M}} = f^{\mathcal{I}}(t_1^{\mathcal{M}}, \ldots, t_n^{\mathcal{M}})$.

We say the closed term t *names*, or *designates* the object $t^{\mathcal{M}} \in \mathcal{D}$.

This definition can be extended to terms containing free variables, but we will not need it until Chapter 7. It can also be extended to an assignment of truth values to closed formulas, something we take up in the next section.

Until we reach Chapter 7 the models we are interested in are rather well-behaved, as far as terms are concerned — they are canonical, in the following sense.

Definition 3.3. A model $\mathcal{M} = \langle \mathcal{D}, \mathcal{I} \rangle$ is *canonical* with respect to a language $L(\mathbf{R}, \mathbf{F}, \mathbf{C})$ if every member of the domain is named by a closed term of the language, that is, if every member of $\mathcal{D}$ is $t^{\mathcal{M}}$ for some closed term t.

Not all models are canonical. For instance, if we set up a language to discuss the real numbers there will be only countably many closed terms, though there are uncountably many reals. Consequently any model whose domain is the reals can not be canonical, no matter how we devise the language.

The machinery introduced in this section is quite general, but in fact we only need to consider a few specific models, the so-called *intended* models for arithmetic and for finite set theory. We introduce them now, beginning with arithmetic. For this, essentially, we want to provide enough machinery to discuss counting, adding, and multiplying. One might reasonably ask, why not also exponentiation, long division with remainder, or other things? As it happens, all these can be defined in a suitable sense, once the basic machinery is present. This will be shown later on. Incidentally, we reserve the use of the usual notation, $+$, $\times$, etc. for the operations themselves, according to the usual mathematical practice. In our formal first-order language we use 'funny' versions, $\oplus$, $\otimes$, etc.

Definition 3.4. *LA* (for 'language of arithmetic') is $L(\mathbf{R}, \mathbf{F}, \mathbf{C})$ where:

1. **R** contains only the binary relation symbol $\approx$;

2. **F** contains the binary function symbols $\oplus$ and $\otimes$ and the unary function symbol $\mathbb{S}$ (successor);

3. **C** contains the symbol **0**.

Naturally the model that goes with this language is the expected one. In fact, it is called *the intended model* or the *standard model* of arithmetic.

Definition 3.5. The model $\mathbb{N}$ has domain ω, and has interpretation $\mathcal{I}$ where:

1. $\approx^{\mathcal{I}}$ is the equality relation, $=$, on ω.

2. $\oplus^{\mathcal{I}}$ is the addition operation, $+$, on ω.

3. $\otimes^{\mathcal{I}}$ is the multiplication operation, $\times$, on ω.

4. $\mathbb{S}^{\mathcal{I}}$ is the operation of adding 1, $x \mapsto x+1$, on ω.

5. $\mathbf{0}^{\mathcal{I}} = 0$.

Generally we will be somewhat informal in our writing of first-order formulas of *LA*. Officially, $\oplus(v_0, \mathbf{0})$ is a term—we will usually write it as $(v_0 \oplus \mathbf{0})$. Indeed, we will feel free to leave out parentheses, or replace them by square or curly brackets, if it will make reading easier. Similarly we will use $\approx$ in infix position, and so on. Also we will generally write x, y, ..., instead of the less easy to read v_0, v_1, ... With this understanding, the 'official' sentence:

$$(\forall v_0)(\forall v_1)((\exists v_2) \approx (\oplus(v_0, v_2), v_1) \vee (\exists v_2) \approx (\oplus(v_1, v_2), v_0))$$

will get abbreviated as:

$$(\forall x)(\forall y)[(\exists z)(x \oplus z \approx y) \vee (\exists z)(y \oplus z \approx x)].$$

Understand, we will do this when we are *using* formulas. When we are *studying* them as mathematical objects, we will use no abbreviations.

Next we turn to the language of set theory. Once again we distinguish between informal mathematical symbols and their formal counterparts by using different alphabets. But there are other issues to be dealt with as well. The model $\mathbb{N}$ is canonical with respect to the language *LA* (see Exercise 3.1). This is a useful feature to have, and we want to carry it over to formal set theory as well. This means that, although set theory is sometimes formalized in a language with only ∈ and no function symbols we can't proceed this way, since we will need closed terms as names.

When function symbols are used in presentations of set theory, they are usually introduced to designate operations like power set, unordered pair, and union. These are natural operations, but there are several of them and for technical purposes it is desirable to keep the number small. It turns out there is a single set-theoretic operation that is also quite natural, though it is less familiar than those just mentioned, and which will allow us to introduce the whole family via definitions, provided it is the hereditarily finite sets we are talking about. It is the operation of adding an additional member to a set. We use the notation $\mathcal{A}(x,y)$ for this operation when we are *doing* mathematics, read it as, "add to the set x the (possibly new) member y", and define it by $\mathcal{A}(x,y) = x \cup \{y\}$.

Definition 3.6. *LS* (for 'language of set theory') is $L(\mathbf{R},\mathbf{F},\mathbf{C})$ where:

1. $\mathbf{R}$ contains the binary relation symbol ε;
2. $\mathbf{F}$ contains the binary function symbol $\mathbb{A}$;
3. $\mathbf{C}$ contains the constant symbol $\varnothing$.

As with arithmetic, we have used "funny" symbols, so we can easily distinguish formal symbols from their informal counterparts. Notice that there is no symbol to represent equality. It turns out we can get the effect of having such a symbol by using the other machinery, so in the interests of simplicity equality is not taken as basic. As we did with *LA*, we will be somewhat informal in formula display, writing ε in infix position. Now, the *intended* or *standard model* for hereditarily finite set theory can probably be guessed at.

Definition 3.7. The model $\mathbb{HF}$ has as domain R_ω, and has interpretation $\mathcal{I}$ where:

1. $\varepsilon^\mathcal{I}$ is the membership relation, $\in$, on R_ω.
2. $\mathbb{A}^\mathcal{I}$ is the add-to operation, $\mathcal{A}$, on R_ω.
3. $\varnothing^\mathcal{I} = \emptyset$.

Exercises

Exercise 3.1. Prove that in the model $\mathbb{N}$, every member of ω is named by infinitely many closed terms of *LA*.

Exercise 3.2. Prove that in the model $\mathbb{HF}$, every member of R_ω except $\emptyset$ is named by infinitely many closed terms of *LS*.

4 Truth in Canonical Models

We have said what structures we are interested in, $\mathbb{N}$ and $\mathbb{HF}$. We have set up languages to talk about these structures, LA and LS. We have even discussed how closed terms name objects in the domains of these structures. But we have not yet said what the *formulas* of these languages are to mean in their respective structures or, more precisely, which of them are to be taken as true.

We owe the definition of truth for sentences of a formal language, in a model, to Tarski. In general the definition is somewhat complicated, owing to the necessity to deal, somehow, with the presence of free variables in formulas. This will be discussed in Chapter 7. But for now we are rather fortunate since both the structures we are interested in are *canonical*: every member of the domain has a name in the language. For canonical models, a definition of truth for sentences can be given that does not require consideration of free variables at all. Consequently, for now we only define truth for sentences in canonical models—it is all we need. We do say, however, what the source of the difficulty is, and then it will be obvious why we can avoid it. The problem arises in trying to give meaning to the quantifiers. Informally, $(\forall x)\varphi(x)$ should be true in a model if $\varphi(x)$ is true of each thing in the model. The trouble is, if there are things in the model that do not have a name in the language we are using, how do we say $\varphi(x)$ is true of them? Tarski's solution is to assign values in the domain of the model directly to variables, like x, whether or not these values have names. Then $(\forall x)\varphi(x)$ is taken to be true if $\varphi(x)$ is true, no matter what value has been assigned to x. This, of course, necessitates that formulas with free variables be dealt with directly, and not just as something whose only purpose is to help us define sentence. We avoid the problem, for now, by simply using closed terms as names and considering only canonical models.

Definition 4.1. Let $L(\mathbf{R}, \mathbf{F}, \mathbf{C})$ be a language and $\mathcal{M} = \langle \mathcal{D}, \mathcal{I} \rangle$ be a canonical model for this language. The atomic sentence $P(t_1, \ldots, t_n)$ is true in $\mathcal{M}$ if the n-tuple $\langle t_1^{\mathcal{M}}, \ldots, t_n^{\mathcal{M}} \rangle$ is in the relation $P^{\mathcal{I}}$.

Example

1. Use the language of arithmetic, and its standard model. The atomic sentence $\mathbb{S}(\mathbf{0}) \oplus \mathbb{S}(\mathbf{0}) \approx \mathbb{S}(\mathbb{S}(\mathbf{0}))$ is true.

2. This time use the language of set theory, and the model $\mathbb{HF}$. Let t and u be any two closed terms. The atomic sentence

$$u \, \varepsilon \, \mathbb{A}(\mathbb{A}(\varnothing, u), t)$$

   is true.

Definition 4.2. Again let $L(\mathbf{R}, \mathbf{F}, \mathbf{C})$ be a language and $\mathcal{M} = \langle \mathcal{D}, \mathcal{I} \rangle$ be a canonical model for this language. Truth for arbitrary sentences of the language L, in the model $\mathcal{M}$, is defined recursively, as follows.

1. Atomic formulas are covered by Definition 4.1.

2. $(A \wedge B)$ is true if A is true and B is true.

3. $(A \vee B)$ is true if A is true or B is true.

4. $(A \supset B)$ is true if A is not true or B is true.

5. $\neg A$ is true if A is not true.

6. $(\forall x)\varphi(x)$ is true if, for every closed term t of the language, $\varphi(t)$ is true.

7. $(\exists x)\varphi(x)$ is true if, for some closed term t of the language, $\varphi(t)$ is true.

Of course, if a sentence is not true, we say it is *false*.

We did not give a truth definition for $\equiv$. We use this binary connective only occasionally, and so rather than taking it as basic, we think of it as defined: $A \equiv B$ should be thought of as abbreviating $(A \supset B) \wedge (B \supset A)$.

Example

1. Using the standard model for arithmetic, $\mathbb{N}$, $(\forall x)(\forall y)(x \oplus y \approx y \oplus x)$ is true, while $(\forall x)(\exists y)(x \otimes y \approx \mathbb{S}(\mathbf{0}))$ is false.

2. Using the standard model for hereditarily finite set theory, $\mathbb{HF}$, $(\forall x)(\exists y)(x \,\varepsilon\, y)$ is true, but $(\forall y)(\exists x)(x \,\varepsilon\, y)$ is false.

Exercises

Exercise 4.1. Which of the following are true in their respective intended models:

1. $\{[(\forall x)(x \,\varepsilon\, t \supset x \,\varepsilon\, u) \wedge (\forall x)(x \,\varepsilon\, u \supset x \,\varepsilon\, t)] \wedge (t \,\varepsilon\, v)\} \supset (u \,\varepsilon\, v)$, where t, u, and v are closed terms.

2. $(\exists x)(x \otimes x \approx \mathbb{S}(\mathbb{S}(\mathbf{0})))$

3. $(\forall x)(\exists y)(\forall z)[(\forall w)(w \,\varepsilon\, z \supset w \,\varepsilon\, x) \supset z \,\varepsilon\, y]$

4. $(\forall x)(\forall y)(\forall z)[(x \oplus y \approx x \oplus z) \supset (y \approx z)]$.

Exercise 4.2. Let $L(\mathbf{R}, \mathbf{F}, \mathbf{C})$ be a language and $\mathcal{M} = \langle \mathcal{D}, \mathcal{I} \rangle$ be a canonical model for this language. Suppose t_1, ..., t_n, u_1, ..., u_n are closed terms in this language and $t_1^{\mathcal{M}} = u_1^{\mathcal{M}}$, ..., $t_n^{\mathcal{M}} = u_n^{\mathcal{M}}$. Let $\varphi(x_1, \ldots, x_n)$ be a formula in the language L, where x_1, ..., x_n are all the free variables, and let $\varphi(t_1, \ldots, t_n)$ and $\varphi(u_1, \ldots, u_n)$ be the result of replacing x_i with t_i and u_i respectively. Show $\varphi(t_1, \ldots, t_n)$ is true in $\mathcal{M}$ if and only if $\varphi(u_1, \ldots, u_n)$ is true in $\mathcal{M}$.

5 Representable Sets and Relations

Statements of LA make assertions about numbers, via the standard model $\mathbb{N}$. Likewise statements of LS make assertions about R_ω. We will see that even though these languages are artificial, the assertions that can be made in them include many of the things that mathematicians need to say: that some member of ω is prime, or that some member of R_ω is a finite sequence, for instance. In fact, we even have the machinery for *defining* the class of primes, or finite sequences, in a straightforward manner. While the notion of definable set or relation is meaningful for any language and any structure for that language, it is easiest to characterize when canonical models are involved, and that is all we talk about now.

Suppose $\varphi(x)$ is a formula of a first-order language L with at most the variable x free. Also let $\mathcal{M}$ be a canonical model for L. For some closed term t the sentence $\varphi(t)$ may be true in $\mathcal{M}$, for another, false. Since every member of the domain of $\mathcal{M}$ is named by some closed term of L, we can think of $\varphi(x)$ as determining a subset of the domain of $\mathcal{M}$, the subset consisting of those things that $\varphi(x)$ is true (of a name) of. Note that by Exercise 4.2 it does not matter which closed term we pick as a name for something, since different terms that name the same thing behave alike in formulas. Now for the formal definition.

Definition 5.1. Let $\mathcal{M} = \langle \mathcal{D}, \mathcal{I} \rangle$ be a canonical model for the language L. With respect to $\mathcal{M}$, the formula $\varphi_1(x)$ of L *represents* the set:

$$\{t^{\mathcal{M}} \mid t \text{ is a closed term of } L \text{ and } \varphi_1(t) \text{ is true in } \mathcal{M}\}.$$

Likewise the formula $\varphi_2(x, y)$ of L *represents* the two-place relation:

$$\{\langle t^{\mathcal{M}}, u^{\mathcal{M}} \rangle \mid t, u \text{ are closed terms of } L \text{ and } \varphi_2(t, u) \text{ is true in } \mathcal{M}\}.$$

Similarly for three-place relations, and so on.

A set or relation is *representable* if some formula represents it.

In general we will omit explicit reference to the language L and the model $\mathcal{M}$, if they are clear from context. In fact, we will only be interested in numbers and hereditarily finite sets.

Example

1. For arithmetic, the formula: $\varphi(x) = (\exists y)(y \oplus \mathbb{S}(\mathbb{S}(\mathbf{0})) \approx x)$ represents the set $\{2, 3, 4, \ldots\}$.

2. Again for arithmetic, the formula: $\varphi(x, y) = (\exists z)(x \oplus z \approx y)$ represents the less-than-or-equal relation, that is, $\{\langle m, n \rangle \mid m \leq n\}$.

3. For set theory, the formula $\varphi(x) = (\exists y)(y \, \varepsilon \, x)$ represents the collection of non-empty sets.

R_ω and ω are both countable, and so have uncountably many subsets. It is not hard to show there are only countably many formulas of LS and of LA. Hence for each structure there are subsets that are not representable by formulas. Loosely, there are subsets that we can not explicitly specify. The natural question is: which subsets do turn out to be representable? We will see some surprising ones are, and some even more surprising ones are not.

Exercises

Exercise 5.1. For arithmetic, show the set of prime numbers is representable.

Exercise 5.2. For set theory, find a formula representing the collection of sets having at least two members.

Exercise 5.3. For set theory, give a formula representing the power set operation: y is the collection of subsets of x.

CHAPTER 3

REPRESENTABILITY FOR SET THEORY

1 Introduction

We want to show several useful and important sets and relations are representable in the structures we are considering, $\mathbb{N}$ and $\mathbb{HF}$, and we want to show the notions of representability for the two structures are closely connected. In fact, they are virtually the same thing. In this chapter we look at representability in $\mathbb{HF}$ only. We will see that a rich family of mathematical notions turns out to be representable. We turn to $\mathbb{N}$ in the following Chapter.

2 When is a formula simple?

We are concerned with which sets are representable using formulas of first-order logic. For both aesthetic and technical reasons, a simple formula is better than a more complicated one. This assumes, of course, that there is some meaningful notion of simplicity. In this section we discuss such a notion—or rather, enough of it for our purposes. We only talk about set theory for now; comparable issues for arithmetic will be dealt with in the next chapter.

Consider the following three sentences of *LS*, where $\varphi(x)$ is some formula whose instances we can check for validity:

1. $(\exists x)(x \,\varepsilon\, t \land \varphi(x))$;

2. $(\exists x)\varphi(x)$;

3. $(\forall x)\varphi(x)$.

Sentence 1 is the simplest in the following sense: it says some value of x makes $\varphi(x)$ true, and it says where to look for the value—look in t. To find an object making $\varphi(x)$ true will involve a search, but it will be a bounded search, analogous to a counting loop in a programming language.

Sentence 2 is less simple. It too says some value of x makes $\varphi(x)$ true, but doesn't tell us where to find one. Nonetheless, if there is such a value we can find it sooner or later. There are only countably many closed terms of

LS; just go through them one by one. Thus this sentence involves a search, but it is an open-ended one. If $(\exists x)\varphi(x)$ is true, by a search through the closed terms we can eventually discover this fact. On the other hand, if it is not true this search will never halt. This time the analogy is with a while loop of a programming language.

Sentence 3 is the most complex of the three. For it to be true, every closed term must make $\varphi(x)$ true, and to check this we need to verify infinitely many facts. Not surprisingly, there is no analog in any programming language.

Definition 2.1. The following formula abbreviations are introduced in LS:

1. $(\exists x \,\varepsilon\, t)\varphi(x)$ for $(\exists x)(x \,\varepsilon\, t \wedge \varphi(x))$;

2. $(\forall x \,\varepsilon\, t)\varphi(x)$ for $(\forall x)(x \,\varepsilon\, t \supset \varphi(x))$.

These are called *bounded quantifiers* (of set theory).

We will try to use bounded quantifiers whenever possible, in an attempt to keep formulas simple. But there is another issue as well—the use of closed terms as names. It is a convenience when showing some set is representable, but it is a minor nuisance when proving something about the family of representing formulas, since it adds another layer to the machinery that must be dealt with. Fortunately the problem can be avoided, since it can be shown that every set that is representable at all is representable using a formula with no function or constant symbols. Now, at last, we are ready for some basic definitions.

Definition 2.2. A formula of LS with no function or constant symbols, and with all its quantifiers bounded, is a Δ_0-*formula*. A subset of R_ω that is represented by some Δ_0 formula is called a Δ_0-*set*. Similarly for n-place relations.

As the discussion above suggests, a Δ_0 set is an entirely constructive thing in the sense that we can *decide* whether something is a member or not, since the truth of a closed Δ_0 formula can be determined in a finite number of steps.

Formulas with unbounded existential quantifiers are next most complicated after Δ_0 formulas. But note, the mere presence of an unbounded existential quantifier doesn't necessarily mean the quantifier will behave existentially. Consider the formula $\neg(\exists x)\neg\varphi(x)$. This acts as if it were $(\forall x)\varphi(x)$—in terms of our earlier discussion it acts like formula 3 and not like formula 2. An existential quantifier inside a negation symbol behaves like a universal quantifier. This also means there is also a problem with quantifiers interacting with implication, by the way. We're interested in formulas in which unbounded existential quantifiers act existentially, never universally. Of course we don't care how bounded quantifiers occur, since these are constructive in their behavior anyway. What we are interested in are those formulas in which the only quantifiers either are bounded, or are unbounded but behave like existential ones. Such things are called Σ formulas.

Definition 2.3. The class of Σ-*formulas* of *LS* is given by the following:

1. Each Δ_0 formula is a Σ formula.

2. If A and B are Σ formulas, so are $A \wedge B$ and $A \vee B$.

3. If A is a Σ formula, so are $(\forall x \, \varepsilon \, y)A$, $(\exists x \, \varepsilon \, y)A$, and $(\exists x)A$, where x and y are variables.

A set represented by a Σ formula is called a Σ-*set*. Similarly for relations.

To give some idea of how all this works, we show a few simple sets and relations are Δ_0; examples of Σ sets that are not Δ_0 will appear in later sections. Here is a short list of Δ_0 formulas that represent important sets or relations, hence showing they are Δ_0 sets and relations. We also say what abbreviation we will be using for each formula. This list will be continued over several sections, and formulas are numbered consecutively throughout, as S-1, S-2, and so on, to suggest they are representations in a set theoretic structure.

Notation Convention We begin a convention that we will follow systematically. An equality symbol was not included as part of *LS*; nonetheless, we will show the equality relation is representable, Δ_0 in fact. When we need the equality relation, we can use the formula that represents it, so it is pretty much like having it built in. It would be convenient for reading purposes to have a simple abbreviation for the formula that represents the equality relation. We don't want to use $x \approx y$, since that makes it hard to distinguish between built-ins and add-ons. Instead we will use the notation x **equals** y, and more generally, we will use san serif bold notation to abbreviate formulas, thus making it easy to recognize when a formula abbreviation is being used.

S-1. The subset relation is represented by the Δ_0 formula (x **subset** y):

$$(\forall z \, \varepsilon \, x)(z \, \varepsilon \, y).$$

S-2. The equality relation is represented by the Δ_0 formula (x equals y):

$$(x \text{ subset } y) \wedge (y \text{ subset } x).$$

S-3. The empty set is represented by the Δ_0 formula (x is $\emptyset$):

$$(\forall y \, \varepsilon \, x) \neg (y \, \varepsilon \, x).$$

S-4. The relation $\{\langle x, y, z \rangle \mid x = \mathcal{A}(y, z)\}$ is represented by the Δ_0 formula x is $\mathbb{A}(y, z)$:

$$(y \text{ subset } x) \wedge (z \, \varepsilon \, x) \wedge (\forall w \, \varepsilon \, x)(w \, \varepsilon \, y \vee w \text{ equals } z).$$

Exercises

Exercise 2.1. Show that every Δ_0 formula is logically equivalent to a Δ_0 formula with no implication symbols, and with all negations at the atomic level—that is, to a formula built up from $(x \, \varepsilon \, y)$ and $\neg(x \, \varepsilon \, y)$ using $\wedge$, $\vee$, and the bounded quantifiers.

3 Formulas as program modules

We show a relation is Δ_0 or Σ by writing a representing formula of the appropriate kind. Writing such a formula has characteristics that are similar to writing a computer program. In fact when looked at from a suitable distance, the two activities are really identical, something we discuss further in Chapter 6. But writing formulas is rather like programming in a fairly low level language—in particular there is no built-in notion of module or procedure. We illustrate the problem using one of the Δ_0 formulas from the previous section.

Being the empty set is Δ_0, and a representing formula is $(\forall y \, \varepsilon \, x) \neg (y \, \varepsilon \, x)$. We introduced an abbreviation for this formula: (x is $\emptyset$), and it serves rather like a subroutine. That is, when writing other formulas, if we want to say something is the empty set we can do so by calling on this formula. But there is a problem with doing this naively. Suppose we want to write a formula $\varphi(x)$ representing $\{\{\emptyset\}\}$; that is, $\varphi(x)$ is to represent the collection whose only member is $\{\emptyset\}$. We can do this by saying x has the empty set as a member, and any member of x is the empty set. Making use of our previous representation for the empty set, we get the following simple formula.

$$\varphi(x) = (\exists y \, \varepsilon \, x)(y \text{ is } \emptyset) \wedge (\forall y \, \varepsilon \, x)(y \text{ is } \emptyset)$$

It seems clear what this is to mean. But if we simply replace the occurrences of x in definition S-3 of (x is $\emptyset$) by occurrences of y to produce (y is $\emptyset$) we get the following unintelligible formula when $\varphi(x)$ is unabbreviated.

$$\varphi(x) = (\exists y \,\varepsilon\, x)(\forall y \,\varepsilon\, y)\neg(y \,\varepsilon\, y) \wedge (\forall y \,\varepsilon\, x)(\forall y \,\varepsilon\, y)\neg(y \,\varepsilon\, y)$$

There is an obvious variable clash here. The free variable x in (x is $\varnothing$) is meant to be like a parameter in a procedure call, while the variable y appearing in the bounded quantifier of the definition of (x is $\varnothing$) is meant to be like a local variable. High level programming languages in effect rename variables so there is no conflict across procedures. Unfortunately first-order logic does not have the machinery for this, so we must do it ourselves.

From now on, when we use a previously specified formula A as part of a new formula B, we will assume all the bound variables of A are renamed to new variables that do not appear anywhere in B outside of A. For example, before using the formula (x is $\varnothing$) in writing $\varphi(x)$, we first rename the bound variable y that occurs in the definition of (x is $\varnothing$) by a new variable, say z, that does not occur in $\varphi(x)$ outside of (x is $\varnothing$). Having done this, the unabbreviated definition for $\varphi(x)$ becomes the following, which is no longer problematic.

$$\varphi(x) = (\exists y \,\varepsilon\, x)(\forall z \,\varepsilon\, y)\neg(z \,\varepsilon\, y) \wedge (\forall y \,\varepsilon\, x)(\forall z \,\varepsilon\, y)\neg(z \,\varepsilon\, y)$$

We assume this bound variable renaming is done 'behind the scenes' and we do not mention it explicitly again. The effect, of course, is that we can freely use old definitions in writing new formulas, treating the old definitions as if they really were self-contained modules or language primitives.

4 Doing Away With Terms

The definitions of Δ_0 and Σ formula specifically do not allow constant or function symbols. In this section we sketch a proof that this is not a serious restriction: the same sets would turn out to be Σ whether constant and function symbols are allowed or not.

The only constant symbol of LS is $\varnothing$, and the only function symbol is $\mathbb{A}$. We have already shown that there are Δ_0 formulas representing the things these are supposed to stand for: the formulas (x is $\varnothing$) and (x is $\mathbb{A}(y,z)$). Now the idea, roughly, is to replace all use of $\varnothing$ and $\mathbb{A}$ by use of these Δ_0 formulas instead. The complexity of the process is due to the fact that terms can have other terms inside them, and those others still. We have to 'unwind' terms.

Let $\varphi(x)$ be a formula of LS that meets the conditions of Definition 2.3 for being a Σ formula, except that it may contain constant or function symbols. We translate it into a proper Σ formula through a sequence of rewritings. The sequence is constructed as follows.

First we eliminate all occurrences of the constant symbol ∅, which is a rather simple process. Pick a variable v that does not occur in $\varphi(x)$ (free or bound). Wherever the symbol ∅ occurs in $\varphi(x)$, replace it with an occurrence of v, and call the result $\varphi_0(x)$. Now, set $\varphi_1(x)$ to be the formula $(\exists v)[(v \text{ is } \emptyset) \wedge \varphi_0(x)]$. This introduces an unbounded existential quantifier, so if we began with a Δ_0 formula we no longer have one. But we do still have a Σ formula (except for the possible occurrences of $\mathbb{A}$). It should be clear that $\varphi_1(x)$ and $\varphi(x)$ represent the same set.

Now suppose there are occurrences of the function symbol $\mathbb{A}$ in $\varphi_1(x)$. Then there must be an innermost such occurrence, of the form $\mathbb{A}(y,z)$ where y and z are variables. Choose such an occurrence; say it occurs as part of the atomic subformula F. Now, pick a variable w that does not occur in $\varphi_1(x)$ and do the following. First, replace the chosen occurrence of $\mathbb{A}(y,z)$ with an occurrence of w; let us say this turns the atomic subformula F into F'. Second, replace the occurrence of F' in $\varphi_1(x)$ with $(\exists w)[(w \text{ is } \mathbb{A}(y,z)) \wedge F']$. Call the resulting formula $\varphi_2(x)$. It is not hard to see that $\varphi_2(x)$ and $\varphi_1(x)$ represent the same set. Note that $\varphi_2(x)$ has one fewer occurrence of a function symbol.

Continuing in this way, produce $\varphi_3(x)$, $\varphi_4(x)$, and so on, until a formula is reached containing no function symbols. This is a proper Σ formula, representing the same set as $\varphi(x)$.

Exercises

Exercise 4.1. Show the set $\{2\}$ is Σ as follows.

1. Write a term t of LS such that $t^{\mathbb{HF}} = 2$.

2. The formula $(x \text{ equals } t)$ represents the set $\{2\}$; apply the procedure outlined above to convert this to a Σ formula.

3. Can it be shown that $\{2\}$ is actually Δ_0?

Exercise 4.2. Show that, for each $s \in R_\omega$, the set $\{s\}$ is Δ_0. Hint: use induction on the rank of s.

5 Some More Δ_0 Sets

In this section we show some basic mathematical notions, like being an ordered pair, or being a function, are Δ_0 in $\mathbb{HF}$. Since members of R_ω are finite, the notion of function we are talking about is that of finite function, of course, but that is quite enough for our purposes. We continue the list of Δ_0 notions begun in Section 2 but we simplify our terminology somewhat. Earlier we spoke of the relation $\{\langle x, y, z\rangle \mid x = \mathcal{A}(y, z)\}$. Now we just talk about the relation $x = \mathcal{A}(y, z)$. Similarly for other relations too, of course.

S-5. The relation $x = \{y, z\}$ is represented by the Δ_0 formula (x is $\{y, z\}$):

$$(y \,\varepsilon\, x) \wedge (z \,\varepsilon\, x) \wedge (\forall v \,\varepsilon\, x)((v \text{ equals } y) \vee (v \text{ equals } z)).$$

S-6. The relation $x = \{y\}$ is represented by the Δ_0 formula (x is $\{y\}$):

$$(x \text{ is } \{y, y\}).$$

S-7. The relation $x = \langle y, z \rangle$ is represented by the Δ_0 formula (x is $\langle y, z \rangle$):

$$(\exists v \,\varepsilon\, x)(v \text{ is } \{y\}) \wedge (\exists v \,\varepsilon\, x)(v \text{ is } \{y, z\}) \wedge \\ (\forall v \,\varepsilon\, x)((v \text{ is } \{y\}) \vee (v \text{ is } \{y, z\}))$$

It is easy to show the set of ordered pairs is Σ; the following formula represents it: $(\exists v)(\exists w)(x \text{ is } \langle v, w \rangle)$. But with only a little more work, this can be improved to Δ_0.

S-8. The subset of R_ω consisting of ordered pairs is represented by the Δ_0 formula Ordpair(x):

$$(\exists u \,\varepsilon\, x)(\exists v \,\varepsilon\, u)(\exists w \,\varepsilon\, u)(x \text{ is } \langle v, w \rangle).$$

The following diagram may help in understanding the Δ_0 formula above. Recall that $\langle v, w \rangle$ is $\{\{v\}, \{v, w\}\}$.

$$\{\{v\}, \underbrace{\{v, w\}}_{u}\}$$
$$\underbrace{\phantom{\{\{v\}, \{v, w\}\}}}_{x}$$

S-9. The relation x is an ordered pair with first component y is represented by the Δ_0 formula (x is $\langle y, \_ \rangle$), and we leave it to you to produce a formula; see Exercise 5.2.

S-10. The relation x is an ordered pair with second component y is represented by a Δ_0 formula (x is $\langle \_, y \rangle$); again see Exercise 5.2.

S-11. The set $\{x \in R_\omega \mid x \text{ is a two-place relation}\}$ is Δ_0, represented by the formula Relation(x), whose writing is left to you; Exercise 5.3.

Notation Convention To make reading simpler, we use

$$(\forall \langle x, y \rangle \,\varepsilon\, z) \varphi(x, y)$$

as an abbreviation for

$$(\forall v\,\varepsilon\,z)(\forall w\,\varepsilon\,v)(\forall x\,\varepsilon\,w)(\forall y\,\varepsilon\,w)\,[(v \text{ is } \langle x,y\rangle) \supset \varphi(x,y)].$$

You should convince yourself that the two have the same meaning. Also notice that if φ is Δ_0, so is $(\forall \langle x,y\rangle\,\varepsilon\,z)\varphi(x,y)$; likewise for Σ. We make a similar abbreviation of

$$(\exists \langle x,y\rangle\,\varepsilon\,z)\varphi(x,y)$$

for

$$(\exists v\,\varepsilon\,z)(\exists w\,\varepsilon\,v)(\exists x\,\varepsilon\,w)(\exists y\,\varepsilon\,w)\,[(v \text{ is } \langle x,y\rangle) \wedge \varphi(x,y)].$$

Now we conclude the section with a few more Δ_0 notions.

S-12. The set $\{x \in R_\omega \mid x \text{ is a function}\}$ is represented by the Δ_0 formula $\mathsf{Function}(x)$:

$$\mathsf{Relation}(x) \wedge (\forall \langle y,z\rangle\,\varepsilon\,x)(\forall \langle y',z'\rangle\,\varepsilon\,x)[(y \text{ equals } y') \supset (z \text{ equals } z')].$$

S-13. The relation: x is the domain of the function y is Δ_0, represented by the Δ_0 formula (x is Domain y). Similarly for x is the range of the function y, represented by (x is Range y); see Exercise 5.4.

Exercises

Exercise 5.1.

1. Prove that a set represented by a formula with two free variables, say $\{\langle x,y\rangle \in R_\omega \mid \varphi(x,y)\}$, is also represented by a formula with one free variable.

2. If $\varphi(x,y)$, in part 1, is a Σ formula, can the one-free-variable formula also be Σ? What about Δ_0?

Exercise 5.2. Give a Δ_0 formula representing the relation: x is an ordered pair with first component y. Similarly for the relation: x is an ordered pair with second component y.

Exercise 5.3. Show $\{x \in R_\omega \mid x \text{ is a two-place relation}\}$ is Δ_0.

Exercise 5.4. Show the following relation is Δ_0: x is the domain of the function y. Similarly for the relation: x is the range of the function y.

Exercise 5.5. Show $(\forall \langle x,y\rangle\,\varepsilon\,z)\varphi(x,y)$ and $(\exists \langle x,y\rangle\,\varepsilon\,z)\varphi(x,y)$ are duals of each other. That is, show $(\forall \langle x,y\rangle\,\varepsilon\,z)\varphi(x,y)$ and $\neg(\exists \langle x,y\rangle\,\varepsilon\,z)\neg\varphi(x,y)$ are equivalent (in $\mathbb{HF}$, for every pair of closed terms t and u substituted for x and y).

6   A Normal Form Theorem

Kleene proved a theorem that said every computable function could be represented in a certain simple, uniform way. This has become known as *Kleene's Normal Form Theorem*. We are going to present a result about Σ sets and relations that has some, though not all, of the characteristics of Kleene's theorem. How this is related to Kleene's formulation will have to wait on further discussion in Section 1 of Chapter 6, and the full Kleene theorem will be given in Section 4 of that Chapter.

Definition 6.1.   A Σ_1 formula of set theory is a formula of the form $(\exists x)\varphi$, where φ is a Δ_0 formula.

Thus a Σ_1 formula is a very special kind of Σ formula. It contains only one unbounded existential quantifier, and that occurs at the outermost level.

Theorem 6.2 (Normal Form Theorem).   *Any relation that is Σ in $\mathbb{HF}$ is represented by some Σ_1 formula.*

Proof   The proof occupies the rest of the section. We present it in a rather informal fashion. Essentially the idea is to show any Σ formula can be converted into a Σ_1 formula while preserving its 'meaning' in the structure $\mathbb{HF}$. So, choose your favorite Σ formula, F, to work with. The first stage in this conversion process for F is to move all unbounded existential quantifiers to the outermost level. We say an existential quantifier is at the outermost level if it does not occur as part of an conjunction, disjunction, or a bounded existential or universal quantifier (thus at most, it is within the scope of other existential quantifiers.) Pick an unbounded existential quantifier in F that is not at the outermost level. We move it outward through a series of small steps, all but one of which are straightforward. We discuss one step in some detail, and sketch others.

Suppose $(\exists x)$ occurs as part of a conjunction in F, say in $A \wedge (\exists x)B$. If x occurs free in A, we can introduce a new variable x', and use it in place of x in $(\exists x)B$, and this preserves meaning. So, without loss of generality, we can assume that in $A \wedge (\exists x)B$, x does not occur free in A. Now, simply replace $A \wedge (\exists x)B$ in F with $(\exists x)(A \wedge B)$. This is justfied by the observation that, if x does not occur free in A, then

$$(A \wedge (\exists x)B) \equiv (\exists x)(A \wedge B)$$

is a valid formula of first-order logic. Other similar transformations are justfied by the following valid formulas.

$$((\exists x)B \wedge A) \equiv (\exists x)(B \wedge A) \quad x \text{ not free in } A$$
$$(A \vee (\exists x)B) \equiv (\exists x)(A \vee B) \quad x \text{ not free in } A$$
$$((\exists x)B \vee A) \equiv (\exists x)(B \vee A) \quad x \text{ not free in } A$$
$$(\exists y \, \varepsilon \, t)(\exists x)A \equiv (\exists x)(\exists y \, \varepsilon \, t)A \quad x \neq y \text{ and } x \text{ not in } t$$

The one remaining case is that of an existential quantifier within a bounded universal quantifier, $(\forall y \, \varepsilon \, t)(\exists x)A$. This is not so easy to deal with, because the equivalence of it and $(\exists x)(\forall y \, \varepsilon \, t)A$ is not a validity of first-order logic. Something specific to the structure of hereditarily finite sets is needed.

Suppose, for simplicity, that $(\forall y \, \varepsilon \, t)(\exists x)A$ is a sentence, and $x \neq y$. Assume it is true in $\mathbb{HF}$. Then, for each $y \in t^{\mathbb{HF}}$ there will be at least one x making A true. Arbitrarily choose one such x for each $y \in t^{\mathbb{HF}}$. The collection of these chosen values for x is a collection of the same size as, or smaller than, the set $t^{\mathbb{HF}}$; so it is a finite collection of hereditarily finite sets, and so *is* a hereditarily finite set. Say it is the hereditarily finite set named by the term s. But then, the sentence $(\forall y \, \varepsilon \, t)(\exists x \, \varepsilon \, s)A$ must also be true. This has introduced an additional term, s, but it can be 'quantified away.' That is, it follows by general principles of first-order logic that (taking z new)

$$(\forall y \, \varepsilon \, t)(\exists x \, \varepsilon \, s)A \supset (\exists z)(\forall y \, \varepsilon \, t)(\exists x \, \varepsilon \, z)A$$

is valid. Putting all this together we have the following: if $(\forall y \, \varepsilon \, t)(\exists x)A$ is true in $\mathbb{HF}$, so is $(\exists z)(\forall y \, \varepsilon \, t)(\exists x \, \varepsilon \, z)A$.

We leave it to you to argue that if $(\exists z)(\forall y \, \varepsilon \, t)(\exists x \, \varepsilon \, z)A$ is true in $\mathbb{HF}$, so is $(\forall y \, \varepsilon \, t)(\exists x)A$. Consequently

$$(\forall y \, \varepsilon \, t)(\exists x)A \equiv (\exists z)(\forall y \, \varepsilon \, t)(\exists x \, \varepsilon \, z)A \quad x \neq y \text{ and } z \text{ new}$$

is true in $\mathbb{HF}$, though it is not valid generally.

Now we have a technique for moving an existential quantifier past a bounded universal quantifier: $(\exists z)(\forall y \, \varepsilon \, w)(\exists x \, \varepsilon \, z)A$ replaces $(\forall y \, \varepsilon \, w)(\exists x)A$.

We are almost done. Starting with the Σ formula F, by a series of moves of the sorts described above, all unbounded existential quantifiers can be lifted to the outermost level. Thus we have converted F into

$$(\exists x_1)(\exists x_2) \cdots (\exists x_n)G$$

where G is Δ_0. All that remains is to collapse the existential quantifiers to a single one, and this can be done easily using ordered pairs. To keep notation simple, we explicitly state the technique for the case where there are just two unbounded existential quantifiers, and describe how to extend it. But this is very easy. Suppose we have the formula:

$$(\exists x_1)(\exists x_2)G.$$

Just replace it with the equivalent formula:

$$(\exists x_3)(\exists w\,\varepsilon\, x_3)(\exists x_1\,\varepsilon\, w)(\exists x_2\,\varepsilon\, w)[x_3 \text{ is } \langle x_1, x_2\rangle \wedge G].$$

If there are more than two unbounded existential quantifiers, collapse them two at a time using the technique above, working from right to left.

This completes the description of how to convert Σ formulas to Σ_1 formulas in $\mathbb{HF}$, and concludes the proof of the Normal Form theorem. ∎

7 Numbers are Δ_0

The set of numbers, ω, is a subset of R_ω. This section is devoted entirely to showing it is Δ_0 in the structure $\mathbb{HF}$. Once we have this out of the way, many useful consequences will follow easily.

Since ω is a collection of hereditarily finite sets, the things we are calling numbers have quite a bit of set-theoretic structure that is rather extraneous to their arithmetic role. But that structure can be used to give very simple, though not immediately intuitively, characterizations of ω, characterizations that make it easy to give Δ_0 representing formulas. The particular characterization we adopt is due to Gödel. We give an informal proof that it does indeed characterize the set of numbers. Our proof is informal because we take for granted several basic facts about ω. Formal arguments can be found in many books on set theory.

Definition 7.1. A set x is *transitive* if the following equivalent conditions hold:

1. $y \in x$ implies $y \subseteq x$;

2. $y \in x$, $z \in y$ imply $z \in x$.

Now we start on a proof that numbers are just transitive sets of transitive sets. Recall from Definition 3.6 of Chapter 1 that $0 = \emptyset$, and $x^+ = x \cup \{x\}$.

Fact 1 Every number is transitive.

Informal Proof We show this by induction. For starters, 0, that is, $\emptyset$, is trivially transitive.

Now suppose n is transitive, and assume $x \in n^+$. We show $x \subseteq n^+$. Since $x \in n^+ = n \cup \{n\}$, either $x \in n$ or $x = n$. Since n is transitive, if $x \in n$ then $x \subseteq n$. So, in either case, $x \subseteq n$. But $n \subseteq n^+$, and this completes the argument. ∎

Fact 2 Every number is a transitive set of transitive sets.

Informal Proof Each number is the set of smaller numbers. ∎

Fact 3 If $x \in R_\omega$ is a transitive set of numbers, then x is a number.

Informal Proof Suppose $x \in R_\omega$ is a transitive set of numbers. x is finite, since it is hereditarily finite. Consequently there must be numbers that are not members of x; let n be the smallest one. That is, n is the smallest number such that $n \notin x$. We claim $x = n$, and so x is a number.

Let $k \in n$. Then k is a smaller number than n, so by definition of n, $k \in x$. Consequently $n \subseteq x$.

Suppose $x \not\subseteq n$; choose some k such that $k \in x$ but $k \notin n$. Since $k \notin n$, either $k = n$ or $n \in k$, by the trichotomy principle. The first of these is impossible, since then $n \in x$, contrary to the definition of n. But if $n \in k$, since we also have $k \in x$, and x is transitive, again $n \in x$, which is impossible. Consequently we must have $x \subseteq n$, and so $x = n$. ∎

Fact 4 If $x \in R_\omega$ is a transitive set of transitive sets, x is a number.

Informal Proof Suppose $x \in R_\omega$ is a transitive set of transitive sets. If we show every *member* of x is a number, we are done by Fact 3.

Suppose some member of x is not a number. Then there must be such a member *of lowest rank*, say c. That is, c is a member of x, c is not a number, and any member of x of lower rank than that of c is a number. We derive a contradiction, thus showing every member of x must be a number.

Now, $c \in x$, and x is transitive, so $c \subseteq x$. Then members of c are also members of x. Also, members of c are of lower rank than that of c. It follows that members of c must be numbers. But c, being a member of x, must be a transitive set. Thus c is a transitive set of numbers, and so is a number by Fact 3, and this is a contradiction. ∎

Main Fact For $x \in R_\omega$, x is a number if and only if x is a transitive set of transitive sets.

With this characterization of ω, it becomes quite easy to give a representing formula that is Δ_0.

S-14. Being transitive is represented by the Δ_0 formula $\mathsf{Transitive}(x)$:

$$(\forall y \; \varepsilon \; x)(\forall z \; \varepsilon \; y)(z \; \varepsilon \; x).$$

S-15. ω is represented by the Δ_0 formula $\mathsf{Number}(x)$:

$$\mathsf{Transitive}(x) \land (\forall y \; \varepsilon \; x)\mathsf{Transitive}(y).$$

S-16. The successor operation on numbers, $y = x^+ = x \cup \{x\}$ is represented by the Δ_0 formula (y is x^+):

$$y \text{ is } \mathbb{A}(x, x).$$

Exercises

Exercise 7.1. Show the two parts of Definition 7.1 are equivalent.

Exercise 7.2. Call $x \in R_\omega$ $\in$-*ordered* if, for every $u, v \in x$: $u \in v$ or $u = v$ or $v \in u$.

1. Prove that no two of $u \in v$, $u = v$, or $v \in u$ can hold. Hint: use the notion of rank.

2. Prove that if if $s \in R_\omega$ is transitive and $\in$-ordered, then s is a member of ω. (The converse is obvious.) Hint: if $s \neq \emptyset$, let n be the largest number in s. Show $n^+ \subseteq s$. If $n^+ \neq s$, let x be a member of $s - n^+$ of lowest rank. Now derive a contradiction.

3. Use the result above to give another Δ_0 characterization of ω.

8 Finite Sequences and Arithmetic

In Section 5 we saw that being a (finite) function was Δ_0, and in Section 7 we saw that the set of numbers was also Δ_0. Then a simple combination of these will give us the machinery of finite sequences. Once we have finite sequences, it is easy to introduce things like addition and multiplication of numbers as well, though here our definitions tend to be Σ rather than Δ_0.

S-17. The relation, x is a finite sequence of length y, is represented by the Δ_0 formula (Sequence x With Domain y):

$$\text{Function}(x) \wedge \text{Number}(y) \wedge (y \text{ is Domain } x).$$

S-18. The set of finite sequences is Σ, and is represented by the formula Sequence(x):

$$(\exists y)(\text{Sequence } x \text{ With Domain } y).$$

S-19. The relation, x is a finite sequence, y is in the domain of x, and the yth term of x is z, is Σ, and is represented by the formula (z is x_y):

$$\text{Sequence}(x) \wedge (\exists w \, \varepsilon \, x)(w \text{ is } \langle y, z \rangle)$$

Notation Convention We introduce some abbreviations that will make formulas easier to read and understand. If $\varphi(x)$ is a formula and s is a finite sequence, we will allow ourselves to write $\varphi(s_n)$, using s_n as if it were a term of *LS*. Formally will take $\varphi(s_n)$ as an abbreviation for the formula $(\exists x)[(x \text{ is } s_n) \wedge \varphi(x)]$. Note that this makes $\varphi(s_n)$ a Σ formula whenever $\varphi(x)$ is. For example, though $(x \text{ is } t_j)$ is an introduced formula, $(s_i \text{ is } t_j)$ is not, but by the convention just described, it abbreviates $(\exists x)[(x \text{ is } s_i) \wedge (x \text{ is } t_j)]$. The convention can be extended to multiple occurrences of term-like expressions—eliminate them one at a time.

Now that we have finite sequences, it is easy to show addition is representable. For example, $8 + 3$ is simply the last term of the sequence $8 + 0$, $8 + 1$, $8 + 2$, $8 + 3$, and this sequence is easy to describe. Its domain is $4 = 3^+$, its 0th term is 8, and each term after the initial one is the numerical successor of the previous term.

S-20. The relation x, y, z are numbers, and $z = x + y$, is represented by the Σ formula $(z \text{ is } x + y)$:

$\text{Number}(x) \wedge \text{Number}(y) \wedge \text{Number}(z) \wedge$
$(\exists w)\{(w \text{ is } y^+) \wedge$
$\quad (\exists s)[(\text{Sequence } s \text{ With Domain } w) \wedge (x \text{ is } s_0) \wedge$
$\quad (\forall n \, \varepsilon \, w)(\forall k \, \varepsilon \, w)[(n \text{ is } k^+) \supset (s_n \text{ is } (s_k)^+)]$
$\quad (z \text{ is } s_y)]\}$

Multiplication is easy to represent, now that we have addition. For example, 8×3 is just the last term of the sequence 8×0, 8×1, 8×2, 8×3, and this is a sequence that starts with 0, and in which each term after the first is its predecessor with 8 added. Exponentiation is dealt with in a similar way.

S-21. The relation x, y, z are numbers, and $z = x \times y$, is represented by a Σ formula $(z \text{ is } x \times y)$, which is left to you; see Exercise 8.3.

S-22. The relation x, y, z are numbers, and $z = x^y$, is represented by a Σ formula $(z \text{ is } x \uparrow y)$, which is left to you; see Exercise 8.7.

Exercises

Exercise 8.1. Show the set of finite sequences is actually Δ_0.

Exercise 8.2. Show the formula $(z \text{ is } x_y)$ can be replaced by an equivalent Δ_0 formula.

Exercise 8.3. Show that multiplication is Σ.

Chapter 3. Representability for Set Theory

Exercise 8.4. Suppose $f : \omega \to \omega$ is a function, and the graph of f is Σ in $\mathbb{HF}$, that is, $\{\langle x, y \rangle \mid f(x) = y\}$ is Σ. Let $a \in \omega$ be a fixed number. Define a function $g : \omega \to \omega$ recursively by setting $g(0) = a$ and $g(n+1) = f(g(n))$. Show that the graph of g is Σ.

Exercise 8.5. Show the following relations are Σ:

1. $z = (x \text{ DIV } y)$, meaning z is the integer quotient on dividing x by y;
2. $z = (x \text{ MOD } y)$, meaning z is the remainder on dividing x by y.

Exercise 8.6. Show the relation $z = (x \text{ BITAND } y)$ is Σ. Hint: see Exercise 5.3. Similarly for $z = (x \text{ BITOR } y)$.

Exercise 8.7. Show that exponentiation is Σ.

Exercise 8.8. Write a formula of LS asserting that addition of numbers is commutative. Caution: (z is $x + y$) abbreviates a Σ formula, but $x + y$ is not a term, and cannot appear as such.

Exercise 8.9. Write a formula of LS asserting that multiplication of numbers is associative.

Exercise 8.10. Give a Σ formula (y is powerset of x) to represent the relation that holds when y is the collection of subsets of x. Hint: Suppose y is the powerset of x. Then the powerset of $x \cup \{a\}$ is $y \cup y'$ where y' is the result of adding a to each member of y.

Exercise 8.11. Suppose we define two new bounded quantifiers: $(\forall x \subseteq y)\varphi$, abbreviating $(\forall x)[(x \subseteq y) \supset \varphi]$ and $(\exists x \subseteq y)\varphi$, abbreviating $(\exists x)[(x \subseteq y) \wedge \varphi]$. Show that if these quantifiers are added to the machinery of Σ-formulas, it is still the case that only Σ relations are representable.

Exercise 8.12. By the *transitive closure* of a set x is meant the *smallest* set y that extends x and is transitive. That is, y is the transitive closure of x if y is transitive, has x as a subset, and there is no proper subset of y that extends x and is transitive. Show the relation y transitive closure of x is Σ.

CHAPTER 4

REPRESENTABILITY FOR ARITHMETIC

1 Basics

In Section 5 of Chapter 2 the general notion of representing a relation by a formula was introduced. Then we spent the entire of Chapter 3 looking at what was representable in the particular structure $\mathbb{HF}$. But we are also interested in the structure $\mathbb{N}$, and this is what the present chapter is about. The general emphasis is somewhat different than in Chapter 3 however. Instead of showing that a variety of useful notions are representable in $\mathbb{N}$ directly by giving explicit definitions, this time we will show that representability in $\mathbb{N}$ and representability in $\mathbb{HF}$ are almost the same thing. Then results about $\mathbb{HF}$ will give us similar results about $\mathbb{N}$ for free.

We have chosen to identify numbers with certain sets, the members of ω. Consequently, when working in $\mathbb{HF}$ we have numbers available, along with much else, including ordered pairs, finite sequences, and so on. In $\mathbb{N}$ the domain is ω exactly—there is nothing else. Also the relations, functions, and constants we are given to work with in $\mathbb{N}$ are quite different than we have in $\mathbb{HF}$. This means we must define notions of Δ_0 and Σ all over, in a way that is appropriate for arithmetic. For one thing, we will allow function and constant symbols to appear. For another, the notion of bounded quantifier must be different.

The relation $x \leq y$ is represented, in the structure $\mathbb{N}$, by the LA formula $(\exists z)(x \oplus z \approx y)$. Likewise $x < y$ is represented by $(\exists z)(\neg(z \approx \mathbf{0}) \wedge (x \oplus z \approx y))$. Using these, we have natural notions of bounded quantification for arithmetic.

Definition 1.1. The following formula abbreviations are introduced in LA:

1. $(\exists x \leq t)\varphi(x)$ for $(\exists x)(x \leq t \wedge \varphi(x))$;

2. $(\forall x \leq t)\varphi(x)$ for $(\forall x)(x \leq t \supset \varphi(x))$;

3. $(\exists x < t)\varphi(x)$ for $(\exists x)(x < t \wedge \varphi(x))$;

4. $(\forall x < t)\varphi(x)$ for $(\forall x)(x < t \supset \varphi(x))$.

These are called *bounded quantifiers of arithmetic*.

Actually, either the pair using $\leq$ or the pair using $<$ is sufficient, because the other kind of bounded quantifier is then definable, as follows.

$$(\forall x < t)\varphi(x) \quad \text{iff} \quad (\forall x \leq t)(x \approx t \vee \varphi(x))$$
$$(\forall x \leq t)\varphi(x) \quad \text{iff} \quad (\forall x < t)\varphi(x) \wedge \varphi(t)$$

And similarly for the existential cases. Now with this one change, definitions of Δ_0 and Σ for arithmetic are pretty much the same as for set theory.

Definition 1.2. A formula of LA with all its quantifiers bounded is a Δ_0-*formula*. A subset of ω that is represented in $\mathbb{N}$ by some Δ_0 formula of LA is called a Δ_0-*set*. Similarly for n-place relations.

Definition 1.3. The class of Σ-*formulas* of LA is given by the following:

1. Each Δ_0 formula is a Σ formula.

2. If A and B are Σ formulas, so are $A \wedge B$ and $A \vee B$.

3. If A is a Σ formula, so are $(\forall x \leq y)A$, $(\exists x \leq y)A$, $(\forall x < y)A$, $(\exists x < y)A$, and $(\exists x)A$, where x and y are variables.

A set represented by a Σ formula is called a Σ-*set*. Similarly for relations.

We are mainly interested in Σ, rather than Δ_0, notions for arithmetic. A major result we are heading for is this: for a set S of numbers, S is Σ in $\mathbb{HF}$ if and only if S is Σ in $\mathbb{N}$. This is an example of what we meant when we said representability for $\mathbb{N}$ and for $\mathbb{HF}$ were essentially the same notions. We have a bit of work to do first, before getting to a proof of this.

2 Arithmetic in Set Theory

It looks like set structures have considerably more machinery than arithmetic, and so it should not be surprising that anything we can do in $\mathbb{N}$ we can also do in $\mathbb{HF}$. In this section we state and prove a precise version of this loose idea.

Theorem 2.1. *Let $\mathcal{R}$ be a relation of numbers, that is, a relation on ω.*

1. *If $\mathcal{R}$ is Σ in $\mathbb{N}$ then $\mathcal{R}$ is Σ in $\mathbb{HF}$.*

2. *If $\mathcal{R}$ is representable in $\mathbb{N}$ then $\mathcal{R}$ is representable in $\mathbb{HF}$.*

Proof We only give the proof for Part 1 since that for Part 2 is essentially the same. To keep notation simple, assume $\mathcal{R}$ is one-place—the general case is similar. Suppose $\mathcal{R}$ is represented in $\mathbb{N}$ by $\varphi(z)$, a Σ formula of arithmetic.

Without loss of generality we can assume that all bounded quantifiers are of the forms $(\forall x < t)$ and $(\exists x < t)$. We turn $\varphi(z)$ into a Σ formula of set theory in a pretty straightforward way.

First we take care of nested terms in the standard way. Rather than describing the algorithm in detail, an example should suffice. Suppose we have the atomic formula:

$$((x \oplus y) \otimes z) \approx (u \oplus v)$$

in which terms are nested. We un-nest them by introducing existential quantifiers to 'hold' intermediate results:

$$(\exists a)(a \approx x \oplus y \wedge \\ (\exists b)(b \approx a \otimes z \wedge \\ (\exists c)(c \approx u \oplus v \wedge \\ b \approx c)))$$

Note that this process introduces existential quantifiers only, so it converts $\varphi(z)$ into another Σ formula, call it $\varphi'(z)$. In it, all function and constant symbols occur only in atomic formulas of the forms:

$$y \approx u \oplus v$$
$$y \approx u \otimes v$$
$$y \approx \mathbb{S}(u)$$
$$y \approx \mathbb{0}$$

Now, replace the atomic formulas of arithmetic in $\varphi'(z)$ with their Σ formula counterparts from set theory as given in Chapter 3:

$(y \text{ is } u + v)$
$(y \text{ is } u \times v)$
$(y \text{ is } u^+)$
$(y \text{ is } \emptyset)$

Call the resulting formula $\varphi''(z)$. The idea, of course, is to replace arithmetic parts by set theoretic parts that behave the same way on members of ω.

The next thing is to take care of bounded quantifiers. But this is easy. Recall that for members of ω, $n < m$ if and only if $n \in m$. So simply replace $(\forall x < y)$ by $(\forall x \, \varepsilon \, y)$, and $(\exists x < y)$ by $(\exists x \, \varepsilon \, y)$. Call the result $\varphi'''(z)$.

Finally we have unbounded existential quantifiers to deal with. We can not leave these alone, because in $\mathbb{N}$ an unbounded quantifier ranges only over numbers, while in $\mathbb{HF}$ it ranges over numbers *and* all other hereditarily finite sets. So the idea is to restrict quantifier ranges, which we do in the most direct way possible. Replace every unbounded universal quantifier, say $(\exists x)\cdots x\cdots$ with $(\exists x)(\mathsf{Number}(x) \wedge \cdots x \cdots)$. Call the result $\varphi''''(z)$. We also have to restrict the unbounded variable z, so finally, let $\varphi'''''(z)$ be $\mathsf{Number}(z) \wedge \varphi''''(z)$.

It is easy to see that $\varphi'''''(z)$ represents, in $\mathbb{HF}$, the same set that $\varphi(z)$ represents in $\mathbb{N}$. ($\varphi'''''(z)$ is a prime example of a simple translation). ∎

3 Gödel's β function

We are heading towards a result that is a converse to the Theorem of the previous section, but we need a few technical items first. The structure $\mathbb{N}$ has addition and multiplication among its given functions. It turns out that we need exponentiation as well; it plays a critical role. We could, of course, just take it as basic along with addition and multiplication, but this is not necessary because Gödel discovered that exponentiation is representable in $\mathbb{N}$, it is Σ in fact. In $\mathbb{HF}$ we have no arithmetic built in, but we saw in Section 8 of Chapter 3 that adding, multiplying, and exponentiation can be introduced once we have the notion of finite sequence available. That is the idea here as well. Gödel's β function is an ingenious device for talking about finite sequences in $\mathbb{N}$, even though we do not have them quite as directly available as we do in $\mathbb{HF}$.

We begin with a few results of number theory—rather ancient results at that. First, something that essentially goes back to Euclid:

- If a and b are any two positive integers, and if d is their greatest common divisor, then d is a linear combination of a and b. That is, there are integers (not necessarily positive), u and v, such that $au + bv = d$.

A proof of this can be found in any elementary book on modern algebra. It can even be shown that d is the smallest positive linear combination of a and b, but this is more than we need. Now, an immediate consequence of this is the following.

- If a and b are relatively prime positive integers, there is a positive integer u such that $au \equiv 1 \pmod{b}$.

This follows quickly from the previous item. If a and b are relatively prime, 1 is their greatest common divisor, so for some u and v, $au + bv = 1$, and so $au \equiv 1 \pmod{b}$. If u is not positive, there are infinitely many positive numbers u' such that $u \equiv u' \pmod{b}$. Choose one; $au' \equiv 1 \pmod{b}$.

Now we come to the key technical item needed, another ancient result from number theory. Recall our convention of using 'number' to mean non-negative integer, i.e. member of ω.

Chapter 4. Representability for Arithmetic

Chinese Remainder Theorem Let $x_0, x_1, \ldots, x_n$ be a sequence of $n+1$ pairwise relatively prime positive integers. Then for any sequence of numbers $k_0, k_1, \ldots, k_n$ of the same length, there is a number z such that, for $i = 0, \ldots, n$, $z \equiv k_i \pmod{x_i}$.

Proof Set $x = x_0 \cdot \ldots \cdot x_n$ and for each $i = 0, \ldots, n$ set $w_i = x_0 \cdot \ldots \cdot x_{i-1} \cdot x_{i+1} \cdot \ldots \cdot x_n$. Then for each i, $x = x_i \cdot w_i$, and x_i and w_i are relatively prime. By the discussion above, there is some positive integer z_i such that $w_i \cdot z_i \equiv 1 \pmod{x_i}$, and so $w_i z_i k_i \equiv k_i \pmod{x_i}$.

Now, set $z = w_0 z_0 k_0 + \cdots w_n z_n k_n$. This easily meets the desired conditions because, using the definition of the w_j,

$$\begin{aligned} z &\equiv w_i z_i k_i \pmod{x_i} \\ &\equiv k_i \pmod{x_i}. \end{aligned}$$

∎

Remember, we are trying to find a Σ way of representing finite sequences in arithmetic. The Chinese Remainder Theorem comes close. Suppose we pick $n+1$ pairwise relatively prime positive integers, $x_0, x_1, \ldots, x_n$, and we hold them fixed. Then for any finite sequence $k_0, k_1, \ldots, k_n$, there is a single number z that, in a sense, encodes the sequence. That is, we can recover the sequence $k_0, k_1, \ldots, k_n$ (up to congruence) by dividing z successively by terms of the fixed sequence $x_0, x_1, \ldots, x_n$, and taking remainders. Now, division and remainder are easy to represent in $\mathbb{N}$; they are Δ_0 in fact. So we almost have an encoding of length $n+1$ finite sequences by numbers (up to congruence, unfortunately). The problem is, we need some way of generating the sequence $x_0, x_1, \ldots, x_n$. The next Theorem, or rather its proof, says there is an easy way of producing a sequence of $n+1$ pairwise relatively prime numbers, starting from any sequence of $n+1$ numbers whatsoever. In stating it and its proof we use the notation $x \mid y$ to mean x divides y. We also write $\text{Rem}(x, y)$ for the remainder on dividing x by y.

Definition 3.1. A function $\beta(z, y, x)$ is defined on ω as follows:

$$\beta(z, y, x) = \text{Rem}\left(z, 1 + (x+1) \cdot y\right).$$

Theorem 3.2. *Let $k_0, k_1, \ldots, k_n$ be any sequence of $n+1$ numbers. There are numbers z and y such that, for each $i = 0, \ldots, n$,*

$$\beta(z, y, i) = k_i.$$

Proof Set $j = \max\{n, k_0, \ldots, k_n\}$, and set $y = j!$. Finally, for $i = 0, \ldots, n$, set $x_i = 1 + (i+1) \cdot y$.

We have a sequence $x_0, x_1, \ldots, x_n$ of numbers, and they are pairwise relatively prime by the following argument. Suppose p is prime, and $p \mid x_i$ and $p \mid x_j$; we derive a contradiction. Assume $i > j$. By the definition of x_i and x_j, $p \mid (i-j) \cdot y$, so either $p \mid (i-j)$ or $p \mid y$. If $p \mid y$, since also $p \mid x_i$, $p \mid 1$, which is impossible since p is a prime. Consequently $p \mid (i-j)$. Now both i and j are $\leq n$, so $0 < i - j \leq n$. But also $p \leq (i-j) \leq n$, so $p \mid n! = y$, and we already showed this was impossible.

Now that we know the x_i sequence is pairwise relatively prime, by the Chinese Remainder Theorem there is a number z such that for $i = 0, \ldots, n$, $z \equiv k_i \pmod{x_i}$. But $k_i \leq j \leq y < x_i$, so in fact $\text{Rem}(z, x_i) = k_i$, that is, $k_1 = \text{Rem}(z, 1 + (i+1) \cdot y) = \beta(z, y, i)$. ∎

The β-function provides a two-parameter coding of finite sequences in arithmetic. For every finite sequence $k_0, k_1, \ldots, k_n$ we can find numbers z and y, which we can think of as jointly encoding the information in the sequence, so that the sequence can be recovered from them as $\beta(z, y, 0), \ldots, \beta(z, y, n)$. The only thing left is to show the β-function is Δ_0, and we leave this to you.

Once the β-function is available, showing that exponentiation is Σ in $\mathbb{N}$ is easy. The number a^b is the last (or the bth member of the sequence a^0, $a^1, a^2, \ldots, a^{b-1}, a^b$, and this is the sequence whose first member is 1, and otherwise each item is a times the one before it. (Incidentally, this approach assigns 0^0 the value 1. 0^0 is a troublesome case which is often left undefined in other areas of mathematics. Taking it to be 1 makes life a little simpler here, and does no harm otherwise.)

Proposition 3.3. *The relation* $c = a^b$ *is Σ in $\mathbb{N}$.*

Proof The β-function is Δ_0. We just write $w = \beta(z, y, x)$ to denote a Δ_0 formula representing it in $\mathbb{N}$. Now using the observations above, $c = a^b$ is true if and only if

$$(\exists z)(\exists y)\Big\{ c = \beta(z, y, b) \wedge$$
$$(\forall i \leq b)(\exists n)\Big[n = \beta(z, y, i) \wedge$$
$$[(i \approx \mathbf{0} \wedge n \approx \mathbb{S}(\mathbf{0})) \vee$$
$$(\exists j \leq i)(\exists m)(i \approx \mathbb{S}(j) \wedge m = \beta(z, y, j) \wedge n \approx a \otimes m)]\Big]\Big\}$$

∎

It should be noted that since Gödel's original work, several alternative and ingenious ways of encoding finite sequences in arithmetic have been developed that avoid the Chinese Remainder Theorem. Any way of encoding finite sequences allows us to represent exponentiation, and finally that is all that matters. Here we have chosen to do things as the master did.

Exercises

Exercise 3.1. Show the Chinese Remainder Theorem can be strengthened to conclude that any two numbers z_1 and z_2 meeting the requirements for z are congruent mod $(x_1 \cdot \ldots \cdot x_n)$.

Exercise 3.2. Show the relation $w = \beta(z, y, x)$ is Δ_0.

4 Set Theory in Arithmetic

In Section 2 we saw that the structure $\mathbb{HF}$ can do what the structure $\mathbb{N}$ can do, as far as representability goes. This is not surprising, since R_ω is a very rich collection, containing many mathematical objects. Now we will see that $\mathbb{N}$ can do what $\mathbb{HF}$ can do, too. This is rather unexpected, at least at first glance, since ω seems fairly uncomplicated. But what Gödel and others discovered is that numbers can encode quite elaborate data structures. In a sense, this accounts for much of the power of modern computers.

Before we get to the technical material, we should say in what sense $\mathbb{N}$ can do what $\mathbb{HF}$ can do, since in $\mathbb{N}$ we only have available a limited selection of hereditarily finite sets: the members of ω. Recall, in Section 5 of Chapter 1 we defined a *Gödel numbering* of the hereditarily finite sets; see Definition 5.3 of Chapter 1. To each hereditarily finite set s is assigned its Gödel number, $\mathcal{G}(s)$, and this mapping from R_ω to ω is one-one and onto. Now, we will show that representability in $\mathbb{HF}$ converts to representability of the corresponding collection of Gödel numbers in $\mathbb{N}$.

Definition 4.1. Let $\mathcal{R}$ be an n-place relation on $\mathbb{HF}$. By $\mathcal{G}(\mathcal{R})$ we mean the n-place relation on ω given by:

$$\mathcal{G}(\mathcal{R}) = \{\langle \mathcal{G}(s_1), \ldots, \mathcal{G}(s_n)\rangle \mid \langle s_1, \ldots, s_n\rangle \in \mathcal{R}\}.$$

Theorem 4.2. *Let $\mathcal{R}$ be a relation of sets, that is, a relation on R_ω.*

1. *If $\mathcal{R}$ is Σ in $\mathbb{HF}$ then $\mathcal{G}(\mathcal{R})$ is Σ in $\mathbb{N}$.*

2. *If $\mathcal{R}$ is representable in $\mathbb{HF}$ then $\mathcal{G}(\mathcal{R})$ is representable in $\mathbb{N}$.*

Proof Part 2 has a proof that is similar to that of Part 1; we omit it. Assume $\mathcal{R}$ is a one-place relation that is Σ in $\mathbb{HF}$; the case where it is many-place is similar. By the Normal Form Theorem, 6.2 in Chapter 3, $\mathcal{R}$ is represented by a Σ_1 formula, $(\exists y)\varphi(y, x)$ where $\varphi(y, x)$ is Δ_0, and by

Chapter 3, Exercise 2.1, we can assume that $\varphi(y,x)$ is built up from $(x \, \varepsilon \, y)$ and $\neg(x \, \varepsilon \, y)$ using $\wedge$, $\vee$, and set-theoretic bounded quantifiers. We convert $(\exists y)\varphi(y,x)$ into a Σ formula in the language LA of arithmetic that acts on Gödel numbers the way $(\exists y)\varphi(y,x)$ acts on the sets the numbers code.

The formula $\varphi(y,x)$ is Δ_0, and so has no function or constant symbols, and all quantifiers are bounded. Our first step is to transform the bounded quantifiers. Now, $(\forall z \, \varepsilon \, t)(\cdots z \cdots)$ abbreviates $(\forall z)(z \, \varepsilon \, t \supset \cdots z \cdots)$. By Exercise 5.5 in Chapter 1, if $s \in t$ then $\mathcal{G}(s) < \mathcal{G}(t)$, so as a hybred intermediate step, replace any subformula of $\varphi(y,x)$ of the form $(\forall z \, \varepsilon \, t)(\cdots z \cdots)$ by one of the form $(\forall z < t)(z \, \varepsilon \, t \supset \cdots z \cdots)$. In a similar way, replace any subformula of the form $(\exists z \, \varepsilon \, t)(\cdots z \cdots)$, which abbreviates $(\exists z)(z \, \varepsilon \, t \wedge \cdots z \cdots)$ with $(\exists z < t)(z \, \varepsilon \, t \wedge \cdots z \cdots)$. Call the resulting formula $\varphi'(y,x)$.

The next step in the translation is based on Proposition 5.4, Chapter 1: for $s, t \in R_\omega$, $s \in t$ if and only if $[\mathcal{G}(t) \text{ DIV } 2^{\mathcal{G}(s)}] \text{ MOD } 2 = 1$. We don't know that this is Δ_0 in $\mathbb{N}$, but fortunately we can manage with something weaker. In Exercise 4.1 you are asked to show it is Σ, and its negation is Σ. Using this, in $\varphi'(y,x)$, replace each atomic subformulas of the form $(s \, \varepsilon \, t)$ by a Σ formula representing $[\mathcal{G}(t) \text{ DIV } 2^{\mathcal{G}(s)}] \text{ MOD } 2 = 1$, and replace each atomic subformula of the form $\neg(s \, \varepsilon \, t)$ by a Σ formula representing the negation of $[\mathcal{G}(t) \text{ DIV } 2^{\mathcal{G}(s)}] \text{ MOD } 2 = 1$. Call the resulting formula $\varphi''(y,x)$.

The formula, $(\exists y)\varphi''(y,x)$ is a formula entirely in the language of arithmetic. Also, it is Σ in the arithmetic sense. And finally, it should be clear that it behaves on Gödel numbers of sets the way $(\exists y)\varphi(y,x)$ behaves on the sets themselves. ∎

Exercises

Exercise 4.1. Show each of the following is Σ in $\mathbb{N}$:

1. $E(s,t)$ which holds when $[\mathcal{G}(t) \text{ DIV } 2^{\mathcal{G}(s)}] \text{ MOD } 2 = 1$.

2. $\overline{E}(s,t)$ which holds when $[\mathcal{G}(t) \text{ DIV } 2^{\mathcal{G}(s)}] \text{ MOD } 2 \neq 1$.

5 Σ is Σ

This section wraps up the relationship between $\mathbb{N}$ and $\mathbb{HF}$ as far as we are concerned. We show the notion of Σ is independent of which structure $\mathbb{N}$ or $\mathbb{HF}$ we choose, and similarly for representability in general. Of course this can't be quite as simple as we just said, because in R_ω we have sets, but in ω we have numbers. Still, there is an isomorphism between these, and this isomorphism carries Σ sets into Σ sets. This is the technically proper version of the results.

Before we formally state and prove the isomorphism theorems just mentioned, we need a simple Lemma. Recall that in Section 5 of Chapter 1 we defined our Gödel numbering, $\mathcal{G} : R_\omega \to \omega$. The definition was pretty simple, so it should not be surprising that we have enough machinery in $\mathbb{HF}$ to define it.

Lemma 5.1. *The relation $y = \mathcal{G}(x)$ is Σ on $\mathbb{HF}$.*

Proof We leave writing the Σ formula to you, after sketching the basic ideas. If x is a set, it can be built up from $\emptyset$ using the operation $\mathcal{A}$. Thus, there is a sequence s of sets in which each item is either $\emptyset$ or is $\mathcal{A}(u,v)$ where u and v are earlier terms of the sequence, and the sequence ends with x. Now, in parallel with the construction of this sequence s we can construct a sequence of the corresponding Gödel numbers, as follows. The Gödel number of $\emptyset$ is 0. And if we know the Gödel numbers of u and v, we can calculate the Gödel number of $\mathcal{A}(u,v)$ using Proposition 5.5 of Chapter 1. In this way we produce a sequence of numbers of the same length as S, ending with y. Now we leave things to you, after helpfully recalling Chapter 3, Exercise 8.6. ■

With this out of the way, things come easily. Essentially there are two results, one concerning sets, one restricted to numbers.

Theorem 5.2. *Let $\mathcal{R}$ be a relation on R_ω.*

1. *$\mathcal{R}$ is Σ in $\mathbb{HF}$ if and only if $\mathcal{G}(\mathcal{R})$ is Σ in $\mathbb{N}$.*

2. *$\mathcal{R}$ is representable in $\mathbb{HF}$ if and only if $\mathcal{G}(\mathcal{R})$ is representable in $\mathbb{N}$.*

Proof We only prove Part 1. And half of this is already done: if $\mathcal{R}$ is Σ in $\mathbb{HF}$ then $\mathcal{G}(\mathcal{R})$ is Σ in $\mathbb{N}$, by Theorem 4.2. In the other direction, suppose $\mathcal{G}(\mathcal{R})$ is Σ in $\mathbb{N}$. Then by Theorem 2.1, $\mathcal{G}(\mathcal{R})$ is also Σ in $\mathbb{HF}$. For convenience, assume $\mathcal{R}$ is a one-place relation, and $\mathcal{G}(\mathcal{R})$ is represented in $\mathbb{HF}$ by the Σ formula $\varphi(x)$. Then $\mathcal{R}$ itself is represented by the formula $\psi(y) =$
$$(\exists x)\,[\varphi(x) \wedge y = \mathcal{G}(x)]$$
where we have used the Σ formula for Gödel numbering from the previous Lemma. ■

Theorem 5.3. *Let $\mathcal{R}$ be a relation on ω.*

1. *$\mathcal{R}$ is Σ in $\mathbb{HF}$ if and only if $\mathcal{R}$ is Σ in $\mathbb{N}$.*

2. *$\mathcal{R}$ is representable in $\mathbb{HF}$ if and only if $\mathcal{R}$ is representable in $\mathbb{N}$.*

We leave the proof of this to you — it is similar to that of the preceeding Theorem. The rather remarkable consequence of all this is that when we are investigating representability, it does not matter whether we work in $\mathbb{N}$ or in $\mathbb{HF}$. The peculiar richness of $\mathbb{HF}$ is only apparent; $\mathbb{N}$ is really $\mathbb{HF}$ in disguise.

In many works authors tend to follow Gödel and show directly that ordered pairs, finite sequences, and the like, can each be coded by numbers in special ways. We have chosen, instead, to show once and for all that we can use the entire machinery of $\mathbb{HF}$ when working with $\mathbb{N}$, and this makes things simpler from here on. This way of relating set-theoretic constructs and integer arithmetic became well-known in the 1960's, partly as a result of several (successful) attempts to generalize the notion of computation in various ways. However it has generally remained part of the folklore of the subject, rather than being treated as central, as we have done here.

Exercises

Exercise 5.1. Prove Theorem 5.3.

CHAPTER 5

TARSKI'S THEOREM OR REPRESENTING REPRESENTABILITY

1 What is a Symbol?

Now that we know representability for the structures of finite set theory and for arithmetic are essentially the same, we can concentrate on one structure with no loss. We choose $\mathbb{HF}$, because its riches are easier to get at. In Chapter 3 we saw that quite a number of useful, basic notions were representable, indeed by Σ formulas. The language LS can express that something is an ordered pair, or a function, or even a number. Now we turn LS on itself, to see if it can express that something is a term, or a formula. We even push this to its extreme, and ask about the representability of the notion of representability itself. This leads directly to Tarski's Theorem, and that in turn to Gödel's Theorem, but these must wait until the necessary background has been developed.

The first question to be faced is an apparently very basic one. How is it even possible for LS to talk about itself, since it talks about sets, not formulas? Well, a formula is made up of symbols from an alphabet. If you go back and look at Section 2 of Chapter 2 you will find that we said: a language, in addition to propositional connectives, etc. contains function symbols, relation symbols, etc. But we never said what mathematical objects symbols were; it hasn't been necessary so far. We did, of course, indicate the typographical signs we would use to denote them, which is not the same thing at all. Now it is time to say what the symbols of LS are. Since we haven't needed to know so far, whatever we say now won't affect what we have done. Obviously we are going to say symbols are sets, and we are going to pick sets that are easy to work with. Beyond that, our choices are quite arbitrary.

It should now be a little clearer how LS can talk about itself. If the alphabet of basic symbols consists of sets, then terms and formulas, being finite sequences of symbols, are finite sequences of sets, and hence are sets. Precisely the subject matter of LS. Now we say which sets these symbols are, in Table 5.1. With this done we can start LS talking about itself.

Variables	v_0 $\langle 0,0 \rangle$	v_1 $\langle 0,1 \rangle$	v_2 $\langle 0,2 \rangle$	$\cdots$ $\cdots$
Connectives	$\neg$ $\langle 1,0 \rangle$	$\wedge$ $\langle 1,1 \rangle$	$\vee$ $\langle 1,2 \rangle$	$\supset$ $\langle 1,3 \rangle$
Quantifiers	$\forall$ $\langle 2,0 \rangle$	$\exists$ $\langle 2,1 \rangle$		
Punctuation	) $\langle 3,0 \rangle$	($\langle 3,1 \rangle$	, $\langle 3,2 \rangle$	
Relation Symbol	ε $\langle 4,0 \rangle$			
Function Symbol	$\mathbb{A}$ $\langle 5,0 \rangle$			
Constant Symbol	$\emptyset$ $\langle 6,0 \rangle$			

Table 5.1. The Symbols of *LS*

S-23. The set of variables of *LS* is Σ, and is represented by the formula Variable(x):

$$(\exists y)(\exists z)[(x \text{ is } \langle y,z \rangle) \wedge (y \text{ is } \emptyset) \wedge \mathsf{Number}(z)].$$

Variables are the only family of symbols that are any work at all, since all other families are finite. We can say something is a quantifier, for instance, by just saying it is $\langle 2,0 \rangle$ or $\langle 2,1 \rangle$. We already know that each one-element subset of R_ω is Δ_0, so we introduce the following convention, which will make reading complicated formulas somewhat simpler.

Notation Convention The expression (x is '$\neg$') abbreviates a Δ_0 formula representing $\{\langle 1,0 \rangle\}$, that is, representing $\{\neg\}$. Likewise (x is '$\forall$') abbreviates a Δ_0 formula representing $\{\langle 2,0 \rangle\}$, that is, representing $\{\forall\}$. And similarly for the other symbols.

Exercises

Exercise 1.1. Show the set of variables is actually Δ_0.

2 Concatenation

Terms and formulas are built up by combining simpler terms and formulas. To do this conveniently, we need some string-manipulating machinery within $\mathbb{HF}$.

Definition 2.1. Let $f, g \in R_\omega$ be finite sequences. By $f * g$ we mean the *concatenation* of f and g, the finite sequence consisting of the items of f followed by the items of g. To be more precise, say the domain of f is the number i, and the domain of g is the number j. Then $f * g$ is the sequence h whose domain is $i + j$, such that for $n < i$, $h(n) = f(n)$, and for $n < j$, $h(i + n) = g(n)$.

We remind you of the Notation Convention from Section 8 of Chapter 3, allowing us to use something like s_n, where s is a finite sequence, as if it were a term for formula writing.

S-24. The relation $h = f * g$ is Σ, and is represented by the formula (h is $f * g$):

$$(\exists i)(\exists j)(\exists k)\{(\text{Sequence } f \text{ With Domain } i) \wedge$$
$$(\text{Sequence } g \text{ With Domain } j) \wedge$$
$$(\text{Sequence } h \text{ With Domain } k) \wedge (k \text{ is } i + j) \wedge$$
$$(\forall n \, \varepsilon \, i)(h_n \text{ is } f_n) \wedge$$
$$(\forall n \, \varepsilon \, j)(\exists m)((m \text{ is } i + n) \wedge (h_m \text{ is } g_n))\}$$

Notation Convention We introduce some more abbreviations, designed to make formulas easier to read.

1. We will allow the use of $f * g$ as if it were a term. That is, if $\varphi(x)$ is a formula, we may write $\varphi(f * g)$, taking it as an abbreviation for:

$$(\exists h)[(h \text{ is } f * g) \wedge \varphi(h)].$$

    Notice that if $\varphi(x)$ is Σ, so is $\varphi(f * g)$.

2. Using the convention above, we write (w is $x * y * z$) as an abbreviation for (w is $(x * y) * z$), that is, for

$$(\exists h)[(h \text{ is } x * y) \wedge (w \text{ is } h * z)].$$

3. Likewise we write (v is $w * x * y * z$) as an abbreviation for (v is $(w * x) * y * z$), and so on.

4. We will use things like $x * y * z$, $w * x * y * z$, ... as terms, just as in item 1 above.

Frequently we will have occasion to adjoin a single additional item to a finite sequence. Fortunately we don't need another lengthy discussion; the machinery for doing this is already at hand. The set $\{\langle 0, s \rangle\}$ is the one-term sequence whose only term is s. If we want to adjoin s to the end of f, we simply concatenate the two sequences f and $\{\langle 0, s \rangle\}$. For reading ease we use $\langle\!\langle s \rangle\!\rangle$ as an abbreviation for $\{\langle 0, s \rangle\}$, and use it as if it were a term of LS.

Notation Convention Yet a few more formula abbreviations.

1. We use $(x$ is $\langle\!\langle s \rangle\!\rangle)$ to abbreviate

$$(\exists z)[(z \text{ is } \varnothing) \wedge (\exists y)(y \text{ is } \langle z, s \rangle \wedge x \text{ is } \{y\})].$$

2. We use $\varphi(\langle\!\langle s \rangle\!\rangle)$ to abbreviate $(\exists x)[x \text{ is } \langle\!\langle s \rangle\!\rangle \wedge \varphi(x)]$.

We should note that our various abbreviations can be combined. For instance, $\varphi(f * \langle\!\langle \text{`)'} \rangle\!\rangle)$ unabbreviates to $(\exists x)[x \text{ is } \langle\!\langle \text{`)'} \rangle\!\rangle \wedge \varphi(f * x)]$, and this in turn unabbreviates as above.

3 Representing Terms

At last we are ready to discuss the representability of the terms of LS, by a formula of LS itself, in the model $\mathbb{HF}$. Although the project may seem a little self-conscious, the idea is really quite elementary. Definition 2.2 of Chapter 2 gives the characterization of term: something built up from the variables and the constant symbols using the function symbols. In order to demonstrate that some expression is a term, the standard way is to produce a suitable *formation sequence for terms*: a sequence t_0, t_1, t_2, ..., t_n such that each member of the sequence is either a variable, or is a constant symbol, or comes from earlier members using a function symbol. A formation sequence shows how terms are built up. So, in order to show t is a term, it comes down to showing there is a formation sequence for terms that has t as a member. We just turn this description into a formula. We remind you that LS has a single constant symbol, $\varnothing$, and a single function symbol, $\mathbb{A}$.

S-25.   The family of formation sequences for terms of LS is represented by the Σ formula $\mathsf{TermSequence}(s)$:

$(\exists d)\{(\text{Sequence } s \text{ With Domain } d) \wedge$
$(\forall n \, \varepsilon \, d)[\mathsf{Variable}(s_n) \vee (s_n \text{ is `}\varnothing\text{'}) \vee$
$(\exists i \, \varepsilon \, n)(\exists j \, \varepsilon \, n)(s_n \text{ is } \langle\!\langle \text{`}\mathbb{A}\text{'} \rangle\!\rangle * \langle\!\langle \text{`(`} \rangle\!\rangle * s_i * \langle\!\langle \text{`,'} \rangle\!\rangle * s_j * \langle\!\langle \text{`)'} \rangle\!\rangle)]\}$

S-26.   The family of terms of LS is represented by the Σ formula $\mathsf{Term}(t)$:

$$(\exists s)[\mathsf{TermSequence}(s) \wedge (\exists n)(t \text{ is } s_n)].$$

We leave it to you, as a simple exercise, to show the following.

S-27.   The family of *closed* terms of LS is represented by a Σ formula, which can be called $\mathsf{ClosedTerm}(t)$.

Exercises

Exercise 3.1. Show the family of closed terms of LS is Σ.

4 Representing Formulas

We have seen that the family of terms of LS is Σ. Showing the family of formulas is Σ can be done in a similar way. We introduce the notion of a *formation sequence for formulas*, just like the version for terms, with a few obvious changes: s is a formation sequence for formulas if each member is atomic, or comes from earlier members using a propositional connective, or comes from an earlier member using a quantifier.

S-28. The family of atomic formulas of LS is represented by the Σ formula $\mathsf{Atomic}(s)$:

$$(\exists t)(\exists u)[\mathsf{Term}(t) \land \mathsf{Term}(u) \land (s \text{ is } \langle\!\langle \text{`('} \rangle\!\rangle * t * \langle\!\langle \text{`}\varepsilon\text{'} \rangle\!\rangle * u * \langle\!\langle \text{`)'} \rangle\!\rangle)].$$

S-29. The family of formation sequences for formulas of LS is represented by the Σ formula $\mathsf{FormulaSequence}(s)$:

$$(\exists d)\{(\text{Sequence } s \text{ With Domain } d) \land$$
$$(\forall n \, \varepsilon \, d)[\mathsf{Atomic}(s_n) \lor$$
$$(\exists i \, \varepsilon \, n)(s_n \text{ is } \langle\!\langle \text{`}\neg\text{'} \rangle\!\rangle * s_i) \lor$$
$$(\exists i \, \varepsilon \, n)(\exists j \, \varepsilon \, n)$$
$$(s_n \text{ is } \langle\!\langle \text{`('} \rangle\!\rangle * s_i * \langle\!\langle \text{`}\land\text{'} \rangle\!\rangle * s_j * \langle\!\langle \text{`)'} \rangle\!\rangle) \lor$$
$$s_n \text{ is } \langle\!\langle \text{`('} \rangle\!\rangle * s_i * \langle\!\langle \text{`}\lor\text{'} \rangle\!\rangle * s_j * \langle\!\langle \text{`)'} \rangle\!\rangle \lor$$
$$s_n \text{ is } \langle\!\langle \text{`('} \rangle\!\rangle * s_i * \langle\!\langle \text{`}\supset\text{'} \rangle\!\rangle * s_j * \langle\!\langle \text{`)'} \rangle\!\rangle) \lor$$
$$(\exists i \, \varepsilon \, n)(\exists v)(\mathsf{Variable}(v) \land$$
$$(s_n \text{ is } \langle\!\langle \text{`('} \rangle\!\rangle * \langle\!\langle \text{`}\forall\text{'} \rangle\!\rangle * v * \langle\!\langle \text{`)'} \rangle\!\rangle * s_i \lor$$
$$s_n \text{ is } \langle\!\langle \text{`('} \rangle\!\rangle * \langle\!\langle \text{`}\exists\text{'} \rangle\!\rangle * v * \langle\!\langle \text{`)'} \rangle\!\rangle * s_i))]\}$$

S-30. The family of formulas of LS is Σ and is represented by the formula $\mathsf{Formula}(f)$:

$$(\exists s)[\mathsf{FormulaSequence}(s) \land (\exists n)(f \text{ is } s_n)].$$

We also need to know which variables occur free in a formula. In Definition 2.7, Chapter 2, we said these were the ones changed by substitution. This is not convenient now, because if we want to say a formula is closed, we must say every variable does not occur free in it, and this involves a universal quantifier. There is a way around the problem, allowing us to use a bounded universal quantifier, but we prefer a different approach which we leave to you.

S-31. The following relation is Σ: f is a formula and s is the set of free variables of f. It is represented by the formula (Formula f With Free s). See Exercise 4.1.

Exercises

Exercise 4.1. Give a Σ formula (Formula f With Free s) representing the relation: f is a formula and s is the set of free variables of f. Hint: first do Exercise 2.3.

5 Substitution

We want to show the substitution of a term for a free variable in a formula is a Σ notion. We begin with substitution in terms, where the clutter is somewhat less and the ideas are easier to understand. One way to carry out a substitution of a term t for a variable v in another term s is this. Construct a formation sequence f for the term s and as you do, construct another sequence g in parallel, just like the first but with t written in place of v. More precisely, if the sequence f has a constant symbol or a variable other than v as its nth item, give g the same n^{th} item. If f has v as its nth item, give g the term t as its nth item. And otherwise, if the nth item of f is produced from two earlier items using a function symbol, let the nth item of g be produced from the corresponding earlier items using the same function symbol. In this way, each item of g will be the same as the corresponding item of f except that it has t in place of v. Now, s is some item of f; the corresponding item of g will be the result of carrying out the substitution of t for v in s. All we have to do is make this informal description into a Σ formula.

S-32. The relation: v is a variable, s, t, and u are terms, and u is the result of substituting t for v in s, is Σ. It is represented by the Σ formula (u is $s\begin{bmatrix}v\\t\end{bmatrix}$):

$$\begin{aligned}
&\textsf{Variable}(v) \land \textsf{Term}(s) \land \textsf{Term}(t) \land \textsf{Term}(u) \land \\
&(\exists f)(\exists g)(\exists n)\{ \\
&\quad (\textsf{Sequence } f \textsf{ With Domain } n) \land \\
&\quad (\textsf{Sequence } g \textsf{ With Domain } n) \land \\
&\quad (\forall k \,\varepsilon\, n)[\\
&\quad\quad ((f_k \textsf{ is } \text{'}\varnothing\text{'}) \land (g_k \textsf{ is } \varnothing)) \lor \\
&\quad\quad (\textsf{Variable}(f_k) \land (v \textsf{ is } f_k) \land (t \textsf{ is } g_k)) \lor \\
&\quad\quad (\textsf{Variable}(f_k) \land \neg(v \textsf{ is } f_k) \land (f_k \textsf{ is } g_k)) \lor \\
&\quad\quad (\exists i\,\varepsilon\, k)(\exists j\,\varepsilon\, k) \\
&\quad\quad\quad (f_k \textsf{ is } \langle\!\langle\text{'}A\text{'}\rangle\!\rangle * \langle\!\langle\text{'}(\text{'}\rangle\!\rangle * f_i * \langle\!\langle\text{',}\text{'}\rangle\!\rangle * f_j * \langle\!\langle\text{'})\text{'}\rangle\!\rangle \land \\
&\quad\quad\quad\; g_k \textsf{ is } \langle\!\langle\text{'}A\text{'}\rangle\!\rangle * \langle\!\langle\text{'}(\text{'}\rangle\!\rangle * g_i * \langle\!\langle\text{',}\text{'}\rangle\!\rangle * g_j * \langle\!\langle\text{'})\text{'}\rangle\!\rangle)] \land \\
&\quad (\exists m)(s \textsf{ is } f_m \land u \textsf{ is } g_m)\}
\end{aligned}$$

Now that terms are out of the way, we turn to formulas. Substitution in atomic formulas is easy—there is only one kind of atomic formula.

S-33. The relation: v is a variable, t is a term, A and B are atomic formulas, and B is the result of substituting t for v in A, is Σ. We continue using our substitution notation: (B is $A\begin{bmatrix}v\\t\end{bmatrix}$) abbreviates the formula below. In the formula, the substitution notation that appears refers to the previous formula, S-32, involving substitution in terms, so there should be no confusion.

$$\mathsf{Atomic}(A) \wedge \mathsf{Atomic}(B) \wedge \mathsf{Variable}(v) \wedge \mathsf{Term}(t) \wedge$$
$$(\exists x)(\exists y)(\exists z)(\exists w)[$$
$$(A \text{ is } \langle\!\langle \text{`(`}\rangle\!\rangle * x * \langle\!\langle \text{`}\varepsilon\text{`}\rangle\!\rangle * y * \langle\!\langle \text{`)`}\rangle\!\rangle) \wedge$$
$$(B \text{ is } \langle\!\langle \text{`(`}\rangle\!\rangle * z * \langle\!\langle \text{`}\varepsilon\text{`}\rangle\!\rangle * w * \langle\!\langle \text{`)`}\rangle\!\rangle) \wedge$$
$$(z \text{ is } x\begin{bmatrix}v\\t\end{bmatrix}) \wedge (w \text{ is } y\begin{bmatrix}v\\t\end{bmatrix})]$$

Finally we come to the big one — substitution in formulas generally. The idea is to use the device of parallel formation sequences, as we did for terms above, but with the conditions given in Definition 2.6, Chapter 2. But, we feel you will find it easier to write a Σ characterization yourself instead of reading one of ours, and it is certainly better for you. So we have made it an exercise.

S-34. The relation: v is a variable, t is a term, X and Y are formulas, and Y is the result of substituting t for free occurrences of v in X, is Σ, and is represented by the formula (Y is $X\begin{bmatrix}v\\t\end{bmatrix}$).

Exercises

Exercise 5.1. Show the relation: v is a variable, t is a term, X and Y are formulas, and Y is the result of substituting t for free occurrences of v in X, is Σ.

6 Representing Representability

In the last few sections we started LS talking about itself in $\mathbb{HF}$, and found it could say quite a bit. It can say something is a term, or a formula, or a sentence. In fact, most of the things we said about LS syntax in earlier chapters turn out to be sayable within LS. But there is also semantics. We have said such and such a set is representable by a formula of LS, and now we want to see if such things can be said within LS. We will be investigating the representabiliy of the very notion of representability. More specifically, is the following set representable:

$$\{\langle x, y\rangle \in R_\omega \mid x \text{ is a formula and } y \text{ is in the set which } x \text{ represents}\}.$$

In fact, it is not! And we will find an even simpler and more dramatic example of a non-representable set.

We begin by asking what it means for the hereditarily finite set s to be in the set represented by $\varphi(x)$. According to our definition it means that $\varphi(t)$ is true in $\mathbb{HF}$, where t is a closed term that names s, that is, $t^{\mathbb{HF}} = s$. Then as a first step in attempting to represent representability, we must represent this notion of a term naming a set. Let us examine what this actually means.

Since $\mathbb{HF}$ is canonical, every member of its domain, R_ω, is named by some closed term. For example, the set $\{\emptyset\}$ is named by the closed term $\mathbb{A}(\emptyset, \emptyset)$, because this closed term names the set $\mathcal{A}(\emptyset, \emptyset) = \emptyset \cup \{\emptyset\} = \{\emptyset\}$. A closed term is a sequence of symbols, and a symbol is itself a set; for instance, $\mathbb{A} = \langle 5, 0 \rangle$. Then, after consulting Table 5.1, we see that the closed term $\mathbb{A}(\emptyset, \emptyset)$ is actually the set:

$$\{\langle 0, \langle 5, 0 \rangle \rangle, \langle 1, \langle 3, 1 \rangle \rangle, \langle 2, \langle 6, 0 \rangle \rangle, \langle 3, \langle 3, 2 \rangle \rangle, \langle 4, \langle 6, 0 \rangle \rangle, \langle 5, \langle 3, 0 \rangle \rangle\}.$$

Thus the assertion that $\mathbb{A}(\emptyset, \emptyset)$ names $\{\emptyset\}$ is really an assertion about a pair of sets. Thus it is at least possible that the naming relation is representable.

In fact, the naming relation is Σ, and we leave it to you to show this. You have already seen the basic idea. If t is a closed term, it is a member of some formation sequence for terms. In parallel with the formation sequence, construct another sequence, whose members are the sets named by the corresponding members of the formation sequence. You will find it rather straightforward (see Exercise 6.1).

S-35. The relation: t is a closed term naming the set s, is Σ, and is represented by the formula Names(t, s).

Notation Convention Representability, as in Definition 5.1, Chapter 2, involved formulas with many free variables. But Exercise 5.1 of Chapter 3 shows that we only need to consider formulas with a single free variable. It is obvious that, by appropriately renaming variables, we can always manage with the same free variable, let us standardize on v_0 for convenience.

Given this convention, the set of representing formulas is just the set with v_0 free.

S-36. The set of representing formulas is Σ, and is represented by the formula RepresentingFormula(f):

$$(\exists x)\{(\text{Formula } f \text{ With Free } x) \wedge (\forall y \, \varepsilon \, x)(y \text{ is `}v_0\text{'})\}.$$

From now on we will almost always be using the substitution operation to replace the variable v_0. Consequently we can introduce a somewhat simplified notation.

S-37. From now on we write:
$$(X \text{ is } Y(t))$$
to abbreviate:
$$(\exists z)\{(z \text{ is } `v_0\text{'}) \wedge X \text{ is } Y\left[\begin{smallmatrix}z\\t\end{smallmatrix}\right]\}.$$

Once again we ask: what does it mean for s to be in the set represented by $\varphi(v_0)$? It means $\varphi(t)$ is *true*, where t names s. We have discussed issues of representing formulas, naming, and substitution. What is left is the big issue, the notion of truth itself. Is the set of sentences of LS true in $\mathbb{HF}$ a representable set? Tarski's Theorem is an answer to this question. Before we address this issue, we turn to a famous paradox of set theory.

Exercises

Exercise 6.1. Show the relation: t is a closed term that names the set s, is Σ.

7 Russell's Paradox

At one time set theory was understood in a naive, informal way. But Bertrand Russell (and independently, E. Zermelo) discovered that this led to a simple but devastating paradox. This paradox will serve to motivate the proof of Tarski's Theorem in the next section.

For the moment we argue informally, using concepts of naive set theory, understanding that some of them may be incorrect. Certainly most sets are not members of themselves. The set of three-member sets does not have three members, so it does not belong to itself, for example. Call a set x *ordinary* if $x \notin x$. It used to be taken as a basic principle of set theory that any property determined a set—the set of things having the property. Being ordinary is a property, so it determines a set: the set of ordinary sets. Call it A. We ask if A itself is ordinary, and we find the following. Because A *is* the collection of ordinary sets,

$$A \text{ is ordinary} \quad \text{iff} \quad A \in A$$

but by definition of ordinary,

$$A \in A \quad \text{iff} \quad A \text{ is not ordinary}$$

and thus

$$A \text{ is ordinary} \quad \text{iff} \quad A \text{ is not ordinary}$$

which is impossible. The conclusion is: A doesn't exist. The further conclusion is: the basic principle that all properties determine sets must be incorrect. Among the consequences of this paradox for set theory were the development of both type theory and axiomatic set theory. For us, an analogous argument will give Tarski's Theorem.

8 Tarski's Theorem

Russell's paradox is about sets, or at least, about our ideas about sets. We have been looking at representable sets, and we begin this section by relocating Russell's argument to this new setting. Suppose $\varphi(v_0)$ is a formula of LS. It represents a certain subset of R_ω; let us write φ_S for this set. Now, φ, being a formula, is itself a set, and so may or may not be in φ_S. For example, if $\varphi(v_0)$ is $\mathsf{Formula}(v_0)$, φ_S is the set of formulas, so φ belongs. We call a formula *ordinary* if it does not belong to the set it represents, just as in the previous section we called a set ordinary if it did not belong to itself. Now it should be clear how Russell's argument can be applied, but first let us make all this official.

Definition 8.1.

1. If $\varphi(v_0)$ is a representing formula, φ_S is the set that it represents.

2. A representing formula $\varphi(v_0)$ is *ordinary* if $\varphi \notin \varphi_S$.

We will show that if the set of ordinary formulas were representable, any formula that represents it would be ordinary if and only if was not. Hence the set of ordinary formulas isn't representable (our first example of a nonrepresentable set). And we will show it would be representable if the set of true sentences of LS were, hence it isn't either.

Lemma A The set of ordinary formulas is not representable.

Proof Suppose otherwise. Let $A(v_0)$ be a formula that represents the set of ordinary formulas, so A_S is the set of ordinary formulas. Then, as with Russell's paradox, because A_S is the set of ordinary formulas,

$$A \text{ is ordinary} \quad \text{iff} \quad A \in A_S$$

but by definition of ordinary,

$$A \in A_S \quad \text{iff} \quad A \text{ is not ordinary}$$

and thus

$$A \text{ is ordinary} \quad \text{iff} \quad A \text{ is not ordinary}$$

and this contradiction establishes that such a formula can not exist. ∎

It follows from this Lemma that representability is not representable (see Exercise 8.1). But we can do better than this.

Lemma B Let $\mathcal{T}$ be the set of sentences of LS that are true in $\mathbb{HF}$. If $\mathcal{T}$ were representable, the set of ordinary formulas would also be representable.

Proof The set of ordinary formulas consists of all formulas $\varphi(v_0)$ such that, for some closed term t naming $\varphi(v_0)$, $\varphi(t)$ is not true. Suppose the formula $T(x)$ represents $\mathcal{T}$. Then the following represents the set of ordinary formulas: $\mathsf{Ordinary}(v_0) =$

$$(\exists x)(\exists t)\{\mathsf{RepresentingFormula}(v_0) \wedge \\ \mathsf{Names}(t, v_0) \wedge (x \text{ is } v_0(t)) \wedge \\ \neg T(x)\}$$

∎

Lemma A and Lemma B together establish a version of Tarski's Theorem for sets—it was originally about numbers and we will prove an arithmetic version in Section 10.

Theorem 8.2 (Tarski's Theorem). *The set of sentences of* LS *that are true in the intended model* $\mathbb{HF}$ *is not representable.*

Exercises

Exercise 8.1. Show the relation: f is a representing formula and s is in the set it represents, is not representable.

9 Liars and Fixed Points

Let us take another look at what went on in the proof of Tarski's Theorem. In our approach we related it to Russell's paradox, but it is close kin to a much older paradox as well, the liar paradox. First, let us introduce some useful notation.

Definition 9.1. If s is an hereditarily finite set, $\ulcorner s \urcorner$ is some closed term of LS naming s. If we need to be more definite, we can take $\ulcorner s \urcorner$ to be the first closed term that names s in some standard enumeration of closed terms.

As a matter of fact, it will not matter for our purposes which of the many closed terms naming s we take $\ulcorner s \urcorner$ to be. In this chapter we are considering truth in the standard model, and if closed terms t and u name the same set, $\varphi(t) \equiv \varphi(u)$ is true in $\mathbb{HF}$ for any formula $\varphi(x)$. In later chapters we will consider provability in various formal theories, but we will only look at those theories in which $\varphi(t) \equiv \varphi(u)$ is provable if t and u name the same set. So, the ambiguity in the notation $\ulcorner s \urcorner$ is harmless.

A representing formula X, or more elaborately, $X(v_0)$, is ordinary if and only if $X \notin X_S$, and this means $X(\ulcorner X \urcorner)$ is false. Suppose $A(v_0)$ represented the set of ordinary formulas. Then

$$A(\ulcorner X \urcorner) \text{ 'says' } X(\ulcorner X \urcorner) \text{ is false.}$$

This is the case for any X, so in particular we can take X to be A. Then

$$A(\ulcorner A \urcorner) \text{ 'says' } A(\ulcorner A \urcorner) \text{ is false.}$$

In other words, $A(\ulcorner A \urcorner)$ asserts its own falsehood.

There is a famous paradox, the liar, dating back to the ancient Greeks (at least). Consider the following:

$$\text{This sentence is false.}$$

It is easy to see that the sentence above is true if and only if it is false. This is impossible. One conclusion that might be drawn is that the expression is not a proper sentence, despite its reasonable grammatical appearance. If one adopts this conclusion, it follows that the notion of sentence in an informal language like English is not a simple, well-understood notion.

In a formal language like LS the notion of sentence is well-defined, and each sentence is either true or false in a model, but never both. $A(\ulcorner A \urcorner)$ asserts its own falsehood, so it would be true if and only if it were false. Our conclusion must be that $A(\ulcorner A \urcorner)$ does not exist. Its existence is immediate if the set of true sentences is representable, so it isn't. The old Greek paradox has been used to produce a result of mathematical significance.

There is yet another way of looking at the argument, involving a fixed point theorem, and this will have interesting consequences in Chapter 10. Gödel's original proof of his first incompleteness theorem involved an implicit fixed point construction. Carnap abstracted a more general version and made it explicit, though it was still established using Gödel's argument. This has become known as the *Gödel Fixed Point Theorem*. It has to do with provability in a formal system, but there is a similar version that applies semantically, and can easily be abstracted from our Russell paradox argument. We do this, then use it to get another proof of Tarski's Theorem (or really, the same one in disguise).

Look at the formula given in the proof of Lemma B in Section 8, to represent the set of ordinary formulas. Or rather, it is not a formula, since it turns out that there is no formula $\neg T(x)$ to represent being not true. Suppose, in that 'pseudo' formula, we replace $\neg T(x)$ with $\varphi(x)$, where this really is a formula. Let us call the resulting formula $A(v_0)$. That is, we are given a formula $\varphi(x)$, and we then define $A(v_0)$ as follows.

$$A(v_0) = (\exists x)(\exists t)\{\mathsf{RepresentingFormula}(v_0) \wedge \\ \mathsf{Names}(t, v_0) \wedge (x \text{ is } v_0(t)) \wedge \\ \varphi(x)\}$$

Unwinding the definition, we see that formula A represents the set of representing formulas G such that $G(\ulcorner G \urcorner)$ is in the set that φ represents. If, as earlier, we let φ_S be the set the formula φ represents, and similarly for A_S, we have the following.

$$G \in A_S \quad \text{iff} \quad G(\ulcorner G \urcorner) \in \varphi_S$$

Then in particular, taking G to be A, we also have

$$A \in A_S \quad \text{iff} \quad A(\ulcorner A \urcorner) \in \varphi_S$$

which can be written as

$$A(\ulcorner A \urcorner) \quad \text{iff} \quad \varphi(\ulcorner A(\ulcorner A \urcorner) \urcorner)$$

or equivalently as the assertion that the following is true.

$$A(\ulcorner A \urcorner) \equiv \varphi(\ulcorner A(\ulcorner A \urcorner) \urcorner)$$

We have now proved the following, in which we write X for $A(\ulcorner A \urcorner)$.

Theorem 9.2 (Semantic Fixed Point Theorem). *If φ is a formula with one free variable, there is a 'fixed point' formula X such that $\varphi(\ulcorner X \urcorner) \equiv X$ is true in $\mathbb{HF}$.*

The Fixed Point Theorem gives us a quick second proof of Tarski's Theorem, or rather a second look at the first proof. Suppose the set $\mathcal{T}$ of sentences of LS that are true in $\mathbb{HF}$ were representable, say by the formula $T(v_0)$. In the Fixed Point Theorem take $\varphi(v_0)$ to be $\neg T(v_0)$. Then there is an X such that $\neg T(\ulcorner X \urcorner) \equiv X$. But also, since $T(v_0)$ represents $\mathcal{T}$, the set of true sentences, we must also have $T(\ulcorner X \urcorner) \equiv X$. But then $\neg T(\ulcorner X \urcorner) \equiv T(\ulcorner X \urcorner)$ which is impossible. Conclusion: there is no such formula $T(v_0)$.

10 Tarski's Theorem, Continued

Tarski proved his Theorem about arithmetic. The version we have proved is about sets. It is not hard to get from our version to the original one, however. One can either transfer the proof, or transfer the result itself. In this section we outline both approaches. There is a small piece of work that must be done first, however.

In Section 1 we declared the symbols of the alphabet of LS to be sets, and it followed that the formulas of LS constituted a Σ subset of R_ω. Well, in exactly the same way we can take the symbols of LA to be sets; details are arbitrary, and we leave them to you. Then the collection of formulas of LA also becomes a Σ subset of R_ω. It is not necessary to go through all this in detail—we simply observe that things like substitution of a term for a variable, and the notion of a closed term of arithmetic naming a number, all turn out to be Σ relations. We assume this in what follows. Also since formulas are members of R_ω, they all have Gödel numbers. Now, we can establish Tarski's Theorem in its original form, by moving the proof from sets to numbers.

Theorem 10.1 (Tarski's Theorem). *The set of Gödel numbers of sentences of* LA *that are true in the intended model* $\mathbb{N}$ *is not representable.*

To prove this directly, we need analogs of Lemmas A and B, and these are quite straightforward. For starters, we modify earlier definitions in the obvious way. We say $\varphi(v_0)$ is a *representing formula of arithmetic* if it is in the language LA, and has at most v_0 free, and we write φ_S for the set it represents, in the structure $\mathbb{N}$. Now we call $\varphi(v_0)$ *ordinary* if $\mathcal{G}(\varphi) \notin \varphi_S$.

Lemma A The set of Gödel numbers of ordinary formulas of arithmetic is not representable in $\mathbb{N}$.

Proof Suppose $A(v_0)$ were a formula of LA that represents the set of Gödel numbers of ordinary formulas. That is,

$$\mathcal{G}(\varphi) \in A_S \text{ if and only if } \varphi \text{ is ordinary.}$$

Now we proceed essentially as before:

$$\begin{aligned} A \text{ is ordinary} \quad &\text{iff} \quad \mathcal{G}(A) \in A_S \\ &\text{iff} \quad A \text{ is not ordinary.} \end{aligned}$$

∎

Lemma B Let $\mathcal{T}$ be the set of Gödel numbers of sentences of LA that are true in $\mathbb{N}$. If $\mathcal{T}$ were representable in $\mathbb{N}$, the set of ordinary formulas would also be representable.

Proof If $\mathcal{T}$ were representable in $\mathbb{N}$, $\mathcal{H}(\mathcal{T})$ would be representable in $\mathbb{HF}$. (Recall, $\mathcal{H}$ is the inverse of the Gödel numbering.) That is, the set of true sentences of arithmetic would be representable in $\mathbb{HF}$. Say it is represented by $T(x)$. Then, just as before, $\mathsf{Ordinary}(v_0) =$

Chapter 5. Tarski's Theorem

$$(\exists x)(\exists t)\{\textsf{RepresentingFormula}(v_0) \land$$
$$\textsf{Names}(t, v_0) \land (x \textsf{ is } v_0(t)) \land$$
$$\neg T(x)\}$$

would represent the ordinary formulas of arithmetic, in $\mathbb{HF}$. (Subformulas have slightly different meanings than before, so RepresentingFormula(v_0) represents the collection of representing formulas in the language LA, and similarly for Names(t, v_0,) and substitution.) But then the collection of Gödel numbers of ordinary formulas would be representable in $\mathbb{N}$, by Theorem 5.2 of Chapter 4. ∎

Lemmas A and B, in these versions, give us Tarski's Theorem for arithmetic. In effect, the set version of the proof has been modified to yield the arithmetic version. As we remarked earlier, it is also possible to transfer the result directly without reworking its proof. This approach is perhaps easier conceptually—we describe the basic idea informally, then make it more rigorous below.

Suppose we know how to test sentences of arithmetic for truth. Then we would have a way of testing sentences of set theory for truth as well, because we have a technique for translating sentences of set theory into sentences of arithmetic. Since we don't have a truth test for set theory, we don't have one for arithmetic either. Now, making this rather simple idea rigorous is fairly straightforward, though the full details are tedious.

In Section 4 of Chapter 4 we showed that representability of a relation in $\mathbb{HF}$ implied representability of the corresponding relation of Gödel numbers, in arithmetic. To show this we gave an informal procedure for converting a formula of LS (without function or constant symbols) into a formula of arithmetic having the same effect on Gödel numbers that the original formula did on sets. In effect, we defined a transformation function from the language LS to the language LA that, in a sense, preserved meaning. Call this transformation Φ. Then, if X is a formula of set theory (without constant or function symbols), $\Phi(X)$ is a formula of arithmetic. Now, the description of the transformation, given in Section 4, Chapter 4 is simple and straightforward, and it should not be surprising that it can be captured by a formula. More precisely, the relation $Y = \Phi(X)$ can be shown to be Σ on $\mathbb{HF}$.

Further, in Section 4 of Chapter 3 we gave an informal algorithm for getting rid of constant and function symbols in formulas of LS. That is, we showed how a formula X of LS could be turned into a formula X' with the same meaning, but with no constant or function symbols. Let us call this transformation Ψ; then if X is a formula of LS, $\Psi(X)$ is a formula with the same meaning but without constant or function symbols. Once again the procedure is straightforward, and it can be shown that the relation $Y = \Psi(X)$ is Σ on $\mathbb{HF}$.

Both of the items above are quite tedious to show rigorously, though neither is surprising. We omit the proofs here. But with these items available it is easy to transfer Tarski's Theorem from set theory to arithmetic. Very simply, suppose the set of Gödel numbers of true sentences of arithmetic were representable in $\mathbb{N}$. Since representability for sets of numbers is the same in $\mathbb{N}$ and in $\mathbb{HF}$, we can assume there is a formula $T(v_0)$ that represents the set of Gödel numbers of true sentences of arithmetic, in $\mathbb{HF}$. Then the following formula would represent the set of true sentences of set theory:

$$(\exists x)(\exists y)[y = \Phi(v_0) \wedge x = \Psi(y) \wedge T(x)].$$

Since the set of true sentences of set theory is not representable, $T(x)$ does not exist.

CHAPTER 6

COMPUTABILITY

1 The Importance of Being Σ

We have seen that the notion of being Σ is remarkably stable. It does not matter whether we are talking about $\mathbb{N}$ or $\mathbb{HF}$, and it does not matter whether we are talking about sets or their Gödel numbers. If a relation is Σ in any of these senses, it is Σ in all of them. At the very least, this says we are dealing with a natural class of relations. But this naturalness is even stronger than it seems at first glance. In this section we informally discuss in what way this is so.

We begin our discussion with $\mathbb{HF}$, and we start with the simpler notion of Δ_0. Suppose φ is a closed instance of a Δ_0 formula; that is, it arose by substituting closed terms for all the free variables of some Δ_0 formula. Such a sentence, φ, makes an assertion that is as concrete as anything in mathematics; briefly, the truth of sentence φ in $\mathbb{HF}$ can be determined in a finite number of steps, by a simple algorithm.

If φ is an atomic sentence, it is of the form $(t \, \varepsilon \, u)$, where t and u are closed terms. A closed term can be thought of as instructions for building a set. Follow the instructions embodied in u, and see if the set t constructs turns up. Of course this is quite loose, but it can be made precise. Briefly, atomic sentences can be checked for truth or falsity.

Beyond the atomic level, we proceed by induction on the complexity of φ. For instance, if $\varphi = \neg \psi$, determine the truth value of ψ, and give the opposite answer for φ. Other propositional connectives are treated in a similar way. If $\varphi = (\forall x \, \varepsilon \, t) \psi(x)$, things are only slightly more complex. The closed term t names some finite set, say s. Each member of s in turn is named by some closed term; say $t_1, t_2, \ldots, t_n$ are closed terms that name the members of s. Then, check each of $\psi(t_1), \psi(t_2), \ldots, \psi(t_n)$ for truth, and if all turn out to be true, so is φ; otherwise φ is false. Similarly for the other cases.

By the discussion above, truth or falsity of an instance of a Δ_0 formula is ascertainable, and even a constructively inclined mathematician would agree.

When it comes to closed instances of Σ formulas, things are a little more complicated: we can check for truth, but not for falsity. Let us be more precise. By the Normal Form Theorem in Section 6 of Chapter 3, it is enough to work with Σ_1 formulas. Say $(\exists x)\varphi(x)$ is a closed instance of a Σ_1 formula, so that $\varphi(x)$ is a Δ_0 formula. The collection of formulas of LS is countable, so in particular we can enumerate closed instances of $\varphi(x)$, say as $\varphi(t_0)$, $\varphi(t_1)$, $\varphi(t_2)$, ... Each of these is a closed instance of a Δ_0 formula, so the truth of each can be tested. Do this with one formula after another. If we ever find an i such that $\varphi(t_i)$ is true, we can stop, and announce that $(\exists x)\varphi(x)$ is true. Otherwise our testing procedure will never terminate. In short, we have an algorithm that will enable us to discover $(\exists x)\varphi(x)$ is true, if it is, and which will not terminate otherwise. Such an algorithm is called a *semi-decision procedure*. More precisely, a semi-decision procedure for membership in a set or relation is an algorithm that will never give incorrect answers (though it may not answer some questions at all), and will report correctly that something is a member of the set when in fact it is.

So far, we have argued that Σ relations are subject to algorithmic methods, in the sense that each has a semi-decision procedure. Of course our discussion has been quite informal, but it should be sufficiently convincing. Now we want to argue that the converse is true as well; any relation on R_ω which has an informal semi-decision procedure will be Σ in a formal sense. This is not something that can be rigorously proved, of course, because we are dealing with an informal notion, and not with something that has a clear and precise definition. The best we can do is call on your experience with computer programs. Most of you have some acquaintance with such things, and are probably familiar with counting loops, while loops, and recursive procedures. In a fairly straightforward manner each of these notions can be captured using the machinery of Σ representabiliy in $\mathbb{HF}$. Counting loops correspond directly to bounded quantification in which the quantifier bounds are terms designating numbers. In a similar way, while loops correspond to unbounded existential quantifiers. The machinery of recursive procedures is a little more elaborate, but not beyond us. The use of recursion in a program can be replaced with a while loop in whose body there is manipulation of a stack. As noted, while loops correspond to existential quantifiers, and a little thought will convince you that R_ω has all that is needed to create stack representations.

In the preceeding paragraph we have argued, quite informally, that Σ representability is strong enough to mimic computer programming. This can be made more precise. Formal models of computers can be defined and the behavior of programs on them rigorously specified. Then the connection between programs and Σ formulas can be properly established. This does not fully justify the claims made above—it depends on whether or not you believe that all intuitively conceived algorithms can be programmed using the kinds of computer languages we are familiar with. Most people believe this is so. Many years ago Church and Turing proposed making a version of this the official characterization of computability. It is almost universally accepted today.

The Church-Turing Thesis, version one A relation $\mathcal{R}$ on numbers has a semi-decision procedure if and only if $\mathcal{R}$ is Σ.

This is not exactly what Church and Turing proposed, but it is equivalent, and is well-suited for our purposes. In fact they were not interested in sets and relations, but rather in functions. But it is a short step from one to the other. Suppose we have a function f from numbers to numbers; under what circumstances would we call it computable? We do not require that f be total, that is, the domain of f may be only part of ω. So, informally, f would be considered computable if we had an algorithm such that, for each number n, gives us the value of $f(n)$ if n is in the domain of f, and runs forever if n is not in the domain of f. How can we relate this idea to what was discussed above?

The *graph* of a function f is the relation $\{\langle x, y \rangle \mid x \in \text{domain}(f) \text{ and } y = f(x)\}$. Now suppose f is intuitively computable. Then we have a semi-decision procedure for the graph of f, along the following lines. Suppose we want to test whether $\langle n, k \rangle$ is in the graph of f. Start the algorithm for f working, to compute $f(n)$. If it ever terminates, and has k as answer, then $\langle n, k \rangle$ is in the graph. So we have a way of discovering the answer is yes if $\langle n, k \rangle$ really is in the graph. On the other hand, if n is not in the domain of f, the algorithm will never terminate, so we only have a *semi*-decision procedure.

But this works the other way around as well. Suppose we have a semi-decision procedure for the graph of f; then we have an algorithm for f as well. To compute $f(n)$, start the semi-decision procedure for the graph working on $\langle n, 0 \rangle$. Start a second copy of the semi-decision procedure working on $\langle n, 1 \rangle$, another copy working on $\langle n, 2 \rangle$, and so on. It is possible to manage all this on a single 'machine' by clever time-sharing. If any of these comes up with a yes answer, we have determined the value of $f(n)$. If none of these comes up yes, the process never terminates.

The Church-Turing Thesis, version two A partial function f from numbers to numbers is computable if and only if the graph of f is Σ representable in $\mathbb{HF}$.

Other terminology is more commonly used here, and we introduce it now, though we will continue talking about Σ representability too.

- A relation on numbers that is Σ representable is called *recursively enumerable*, or *semi-computable*.

- A partial function from numbers to numbers whose graph is Σ representable is called *partial recursive*, or *computable*.

There is one more basic concept yet to be introduced, that of a decision procedure for a set or relation. We have discussed semi-decision procedures. These do half the job. If $s \in \mathcal{R}$ and $\mathcal{R}$ is semi-decidable we can discover that s is a member, but if $s \notin \mathcal{R}$ a semi-decision procedure will tell us nothing. A decision procedure should be able to answer yes or no, depending, and not just yes. But the notion of a decision procedure can be reduced to that of a semi-decision procedure, quite directly.

Suppose $\mathcal{R}$ is a relation on numbers. We write $\overline{\mathcal{R}}$ for the complementary relation; that is, if $\mathcal{R}$ is n-place, $\overline{\mathcal{R}}$ is the collection of all n-tuples of numbers not in $\mathcal{R}$. If we had a decision procedure for $\mathcal{R}$, we would also have one for $\overline{\mathcal{R}}$: just reverse answers. And certainly, a decision procedure is also a semi-decision procedure, so both $\mathcal{R}$ and $\overline{\mathcal{R}}$ would each have semi-decision procedures. But the converse is equally true. Suppose both $\mathcal{R}$ and $\overline{\mathcal{R}}$ have semi-decision procedures, and we want to test whether or not $n \in \mathcal{R}$. Start both semi-decision procedures working on n; the one for $\mathcal{R}$ and the one for $\overline{\mathcal{R}}$. Either n is a member of $\mathcal{R}$ or it isn't, so one of the semi-decision procedures must terminate with a yes answer. Running the two semi-decision procedures together amounts to having a decision procedure. So, we consider a set *decidable* if it and its complement both have semi-decision procedures. Once again, other terminology is standard.

- A relation on numbers such that both it and its complement are Σ representable is called *recursive*.

Terminology like recursively enumerable and recursive are standard in the literature, when talking about numbers. But there is one more piece of notation that is often used when taking a set-oriented approach, as we are.

- A relation (on numbers or sets) whose complement is Σ representable is said to be a Π relation. Alternately, a Π relation is one represented by the negation of a Σ formula.

- A relation that is both Σ and Π is said to be a Δ relation. Thus being Δ is alternate terminology for being a recursive relation.

Every Δ_0 formula is also a Σ formula, and the negation of a Δ_0 formula is again a Δ_0 formula. It follows that every Δ_0 relation is also a Δ relation. The converse is not true, but there is more to be said. If a relation $\mathcal{R}$ is Δ_0, there is a formula $\varphi(x)$ that represents it, and $\neg\varphi(x)$ represents the complement. Giving the single Δ_0 formula is enough. But if $\mathcal{R}$ is Δ, there is one Σ formula $\varphi_1(x)$ representing it, and another Σ formula $\varphi_2(x)$ representing its complement, and there may be no obvious relationship between the two formulas. There is no notion of a Δ formula. (This issue came up earlier, in the proof of Theorem 4.2 of Chapter 4. Take a look at it again.)

We have said, quite informally, why being Σ is significant. A function whose graph is Σ is considered computable; a relation that is Δ is considered to have a decision procedure. This makes a connection between the approach here and *recursion theory* or *computability theory*, as found in many other books. Since this is not a work on recursion theory, we have not established connections between Σ representability and machine computation rigorously. This can be done though, and proofs can be found in many places.

Exercises

Exercise 1.1. Suppose f is a function from numbers to numbers that is total, that is, the domain of f is ω. Show that if the graph of f is Σ, then the graph of f is also Π, and hence is Δ.

Exercise 1.2.

1. Give an informal argument to show that if f is a partial recursive function then the domain of f is recursively enumerable.

2. Give an informal argument to show that if $\mathcal{R}$ is recursively enumerable, it is the domain of a partial recursive function.

2 A Σ set, but don't ask Π

In Chapter 3 we spent a great deal of time showing certain relations were not just representable but were Σ, but at the moment we don't know whether our work was real or an illusion. It is at least conceivable that every set representable by a formula is actually representable by a Σ formula. This would mean much of our work was needlessly hard, and the discussion in the previous section looses some of its force. But as a matter of fact there is a natural example of a set that is representable, but not Σ: the set of false instances of Σ sentences. More precisely, the set of false instances of Σ sentences turns out to be Π, so it is certainly representable, but it is not Σ. It will follow that the set of true instances of Σ sentences is Σ but not

Π, and this gives us an example of a set that is recursively enumerable but not recursive. But even more, we will find ourselves touching on some of Turing's work on the theoretical foundations of computer science.

Tarski's Theorem says the set of true sentences of set theory is not representable in set theory. What we will do is mimic the proof of that theorem, adapting it to Σ formulas. Review Chapter 5, Section 8 for terminology and background.

Definition 2.1. A representing formula $\varphi(v_0)$ is Σ-*ordinary* if it is a Σ formula, and $\varphi \notin \varphi_S$.

Lemma A The set of Σ-ordinary formulas is not representable by a Σ formula.

Proof Suppose $A(v_0)$ were a Σ formula that represented the set of Σ-ordinary formulas. Then:

$$A \text{ is } \Sigma\text{-ordinary} \quad \text{iff} \quad A \in A_S$$
$$\text{iff} \quad A \text{ is not } \Sigma\text{-ordinary}.$$

This contradiction concludes the argument. ∎

Being a Σ formula is easily shown to be Σ representable. We leave this to you, and use the fact in the proof of the following.

Lemma B Let $\mathcal{F}$ be the set of instances of Σ formulas that are false in $\mathbb{HF}$. If $\mathcal{F}$ were representable by a Σ formula, the set of Σ-ordinary formulas would also be.

Proof Say $\mathcal{F}$ is represented by the Σ formula $F(x)$. Then the following represents the set of Σ-ordinary formulas:

$$(\exists x)(\exists t)\{\mathsf{RepresentingFormula}(v_0) \wedge$$
$$\Sigma\text{-}\mathsf{Formula}(v_0) \wedge$$
$$\mathsf{Names}(t, v_0) \wedge (x \text{ is } v_0(t)) \wedge$$
$$F(x)\}$$

Notice that if $F(x)$ is Σ, so is the entire formula above. ∎

The two Lemmas above combine to show the following.

Theorem 2.2. *The set of false instances of Σ formulas is not Σ.*

Being a sentence that is an instance of a Σ formula is Σ representable. We leave this to you in the Exercises.

Corollary 2.3. *The set of true instances of Σ formulas is not Π.*

Proof Suppose the set T of true instances of Σ formulas were Π. Then the complement of T would be represented by a Σ formula, say $\overline{T}(x)$. Then the set of false instances of Σ formulas would be represented by the following Σ formula:

$$\Sigma\text{-instance}(v_0) \wedge \overline{T}(v_0)$$

∎

Exercises

Exercise 2.1. Show the set of Σ formulas is Σ representable.

Exercise 2.2. Show the set of closed instances of Σ formulas is Σ.

3 Σ truth is Σ

In the previous section we showed the set of true instances of Σ formulas is not Π. This is a direct analog of Tarski's Theorem. Now we show that, nonetheless, it is Σ. This establishes that Σ and Π really are different. We discuss other consequences later.

The basic idea is quite simple, and essentially amounts to making formal some of the discussion contained in Section 1. To bring out the ideas more clearly, we begin with Δ_0 formulas. Informally, how would we check a Δ_0 formula for truth? Well, if it is atomic we apply a direct test of an appropriate sort. If it is a conjunction, we test each component. If it is a bounded quantification, we test each of the finitely many cases. And so on. Corresponding to this idea of breaking a Δ_0 formula down, we introduce the notion of a *downward saturated* set, a notion due to Hintikka in a somewhat different context.

Definition 3.1. Let $\mathcal{S}$ be a finite set of sentences of LS. We say $\mathcal{S}$ is Δ_0-*downward saturated* if it meets the following conditions:

1. if the atomic sentence $(t \,\varepsilon\, u)$ is in $\mathcal{S}$ then $(t \,\varepsilon\, u)$ is true in $\mathbb{HF}$;

2. if the negated atomic sentence $\neg(t \,\varepsilon\, u)$ is in $\mathcal{S}$ then $(t \,\varepsilon\, u)$ is false in $\mathbb{HF}$;

3. if $(X \wedge Y)$ is in $\mathcal{S}$ so are both of X and Y;

4. if $\neg(X \wedge Y)$ is in $\mathcal{S}$ so is one of $\neg X$ or $\neg Y$;

5. if $(\forall x \,\varepsilon\, t)\varphi(x)$ is in $\mathcal{S}$, and the closed term t names the set s, then for each $s_i \in s$ there is a term t_i naming s_i, with $\varphi(t_i)$ in $\mathcal{S}$;

6. if $\neg(\forall x \,\varepsilon\, t)\varphi(x)$ is in $\mathcal{S}$, and the closed term t names the set s, then for some $s_i \in s$ there is a term t_i naming s_i, with $\varphi(t_i)$ in $\mathcal{S}$.

If $\mathcal{S}$ is Δ_0-downward saturated, every instance of a Δ_0 formula that it contains must be true in $\mathbb{HF}$. This can be shown by induction on formula complexity. Consequently if X is an instance of a Δ_0 formula, and we can find a Δ_0-downward saturated set containing X, we know X is true. But the converse is also the case: if X is true we can verify this fact, and the set of sentences we must consider to establish the truth of X will constitute a Δ_0-downward saturated set. We summarize this in the following.

Lemma 3.2. *Let X be an instance of a Δ_0 formula of LS. X is true in $\mathbb{HF}$ if and only if $X \in \mathcal{S}$ for some Δ_0-downward saturated set $\mathcal{S}$.*

Now what is needed is a Σ formula to represent the collection of Δ_0-downward saturated sets. It is easier to carry out the construction of such a formula than it is to read someone else's version, so we leave this to you. Assuming this has been done, and calling the resulting formula $\Delta_0\text{-DownSat}(v_0)$, we immediately get the following.

Theorem 3.3. *The set of true instances of Δ_0 formulas is Σ.*

Proof The following Σ formula suffices. $\Delta_0\text{-True}(v_0) =$

$$(\exists s)[\Delta_0\text{-DownSat}(s) \wedge (v_0 \; \varepsilon \; s)].$$

∎

Now extending this to instances of Σ formulas is simple.

Definition 3.4. Let $\mathcal{S}$ be a finite set of sentences of *LS*. We say $\mathcal{S}$ is Σ-*downward saturated* if it meets the following conditions:

1. $\mathcal{S}$ is Δ_0-downward saturated;

2. if $(\exists x)\varphi(x)$ is in $\mathcal{S}$ then so is $\varphi(t)$ for some closed term t.

Just as above, it is simple to check that an instance of a Σ formula is true if and only if it belongs to some Σ-downward saturated set. And the collection of downward saturated sets is Σ representable. Then, just as we did with Δ_0 formulas, we get the following, whose obvious proof we omit.

Theorem 3.5. *The set of true instances of Σ formulas is Σ.*

Combining this with Corollary 2.3, we get the following.

Theorem 3.6. *The set of instances of Σ formulas of LS that are true in $\mathbb{HF}$ is Σ but not Δ.*

If we work with Gödel numbers for formulas, instead of the formulas themselves, this becomes Post's Theorem—recall, a set of numbers that is Σ is called recursively enumerable, and one that is Δ is recursive.

Corollary 3.7 (Post's Theorem). *There is a set that is recursively enumerable but not recursive. In particular, the set of Gödel numbers of true instances of Σ formulas of LS is recursively enumerable, but not recursive.*

Note that this means there are sets—important ones at that—that have semi-decision procedures but no decision procedures.

Exercises

Exercise 3.1. Give a Σ formula for the collection of Δ_0-downward saturated sets.

Exercise 3.2. Give a Σ formula for the collection of Σ-downward saturated sets.

Exercise 3.3. Show the set of Gödel numbers of true instances of Σ formulas of arithmetic is recursively enumerable but not recursive.

4 Kleene's Normal Form Theorem

Truth is not representable, but truth for instances of Σ formulas is, by a Σ formula in fact. Representability itself is not representable (Exercise 8.1), Chapter 5. But representability by Σ formulas is representable by a Σ formula. This peculiar fact is the heart of Kleene's Normal Form Theorem, and looked at properly it says there is a universal Turing machine.

Lemma 4.1. *Let $\mathcal{R}$ be the relation: f is a representing formula that is Σ, and s is in the set it represents. $\mathcal{R}$ is Σ.*

Proof $\mathcal{R}(f,s)$ is represented by the Σ formula:

$$\begin{array}{c} \mathsf{RepresentingFormula}(f) \wedge \\ \Sigma\text{-formula}(f) \wedge \\ (\exists t)[\mathsf{Names}(t,s) \wedge \\ (\exists x)(x \text{ is } f(t) \wedge \Sigma\text{-true}(x))] \end{array}$$

where Σ-true is the formula arising from Theorem 3.5. ∎

In Chapter 3, Section 6 we showed that any Σ relation was also Σ_1, that is, it can be represented by a formula with a single initial existential quantifier, governing a Δ_0 formula. Combining this with the Lemma above, we get that there is a Σ_1 formula representing Σ-representability. The following merely states this formally. It is a version of Kleene's Normal Form Theorem. As originally stated, his result had to do with computable functions and arithmetic; the present version is for Σ relations, equivalently, for recursively enumerable relations.

Theorem 4.2 (Kleene's Normal Form Theorem). *There is a formula* $(\exists x)T(x,y,z)$, *where* $T(x,y,z)$ *is* Δ_0, *that indexes the family of* Σ *sets. That is, for each* Σ *set* $\mathcal{P}$ *there is a closed term* f, *called an* index *for* $\mathcal{P}$, *such that* $\mathcal{P}$ *is the set represented by* $(\exists x)T(x,f,v_0)$.

The consequences of this are rather remarkable. Recall, the Σ sets are those that have semi-decision procedures, or are those that can be generated by computer programs. The Kleene Normal Form Theorem says there is a single computer program, the one corresponding to $(\exists x)T(x,y,z)$, that is *universal*. It can represent any Σ set by fixing a parameter, called an *index* above. We can think of $(\exists x)T(x,y,z)$ as like a compiler for some programming language, C say, combined with an operating system, and we can think of indexes as representing C programs. The single compiler/operating system program is then universal in the sense above: by supplying it with a particular C program we get the behavior of that program. And of course, a compiler for C and an operating system can itself be written in C. The Kleene Normal Form Theorem then is an abstract embodiement of what it means to be a universal programming language.

The Normal Form Theorem says more than this, though. Not only is there a universal Σ formula for all Σ sets, but it is of a particularly simple form. It contains a single unbounded existential quantifier; all other quantifiers are bounded. Recall that a bounded quantifier corresponds to a counting loop; an unbounded one to a while loop. Thus the Theorem is essentially asserting that in programming to generate sets, at most a single while loop is needed, and all other loops can be counting loops. This has been established more concretely for particular programming languages.

Finally, use of the Normal Form Theorem gives us a simple, direct proof of the existence of a Σ set that is not Δ, Theorem 3.6. (Actually it is the same proof in disguise.)

Let K be the set represented by the formula $(\exists x)T(x,v_0,v_0)$. Since this is a Σ formula, K is a Σ set. On the other hand, if K were a Δ set, its complement, $\overline{K}$, would be Σ, so by the Kleene Normal Form Theorem $\overline{K}$ would have an index, say f. Then for any set s_0,

$$s_0 \in \overline{K} \text{ iff } (\exists x)T(x,f,s) \text{ where } s \text{ names } s_0.$$

But f names a set, say f_0. Then

$$f_0 \in \overline{K} \Leftrightarrow (\exists x)T(x,f,f)$$

But also by definition of K,

$$(\exists x)T(x,f,f) \Leftrightarrow f_0 \in K$$

and this is a contradiction, establishing that $\overline{K}$ can not be Σ.

This is as far as we take things here. This is not a book on recursion theory or computability theory. We have established a relationship with the subject, and explained why the notion of Σ is as important as it is. We return to our main subject matter after setting things up for you to make connections with the standard literature in the next section.

5 A Turing Machine Sketch

Turing machines are the oldest formal models of what are now called digital computers. The Turing machine concept actually predates the development of the electronic digital computer by a number of years. Turing machines come in several variants—we will describe one particular version. Turing machines can be programmed to compute *functions*: given an input, an output is produced. Or they can be programmed to be *language acceptors*: given a word, it is accepted or not. Here we will describe the use of a Turing machine as a language acceptor.

Think of a Turing machine as having an infinite tape—infinite to the right, with a fixed left end. The tape is ruled into squares, and in each square exactly one symbol can be placed. For convenience we may speak of the symbol in square 0, in square 1, and so on, but understand, this numbering of squares is not part of the formal machinery.

We will have an alphabet with just three symbols, '0', '1', and a blank, which we denote as 'b'. (Other choices of symbols are possible, and are common in automata theory.) When a *run* of a Turing machine begins, we assume some finite sequence of 0's and 1's has been written starting at the left end of the tape, and otherwise the tape is all blanks.

Associated with each Turing machine is a finite set of *states*. Think of the set of states as a small short-term memory. At any given moment a Turing machine is in exactly one state. We assume one state is designated as *start*.

Part of the machinery of a Turing machine is a *read/write head*. This scans a single square of the tape at a time. It "sees" what is on the square, and can write a symbol, and then move left or right one square at a time. Here is a picture to have in mind.

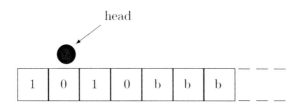

state = s_2

Here the head is scanning square number 1 (start numbering with 0 at the left), sees the symbol 0, and is in state s_2. The tape contains 1010 followed by blanks.

A *program* for a Turing machine consists of instructions of the following two kinds.

Write and Move If the head sees symbol x and is in state s, the head should write symbol x', move one square left or right, and go into state s'.

Halt If the head sees symbol x and is in state s, stop.

One of the symbols that can be written is a blank, so in effect a Turing machine can erase. An instruction of the second kind above is called a *halting instruction*. There must be at least one halting instruction in a program.

We indicate a non-halting instruction using the following abbreviated notation, which we present via an example:

$$0, s_2 \to 1, s_3, L$$

Think of this as the instruction: if 0 is scanned and the machine is in state s_2, replace the 0 by a 1, go into state s_3, and move left one square. Likewise we abbreviate a halting instruction as follows, again presenting a representative example:

$$1, s_3 \to \text{halt}$$

Think of this as the instruction: if 1 is scanned and the machine is in state s_3, halt.

For each instruction we call the left side of the arrow the *conditions* and the right side the *action*. We say a condition is *satisfied* if the head is scanning the symbol indicated, and the machine is in the state given by the condition. We say an action is *executed* if the original symbol is replaced by the new one indicated, the state changes to the new state, and the head moves left or right as indicated. For a halting instruction, executing the action means stopping the machine. Given a set of instructions S and a particular word w written on the tape, a *run* of a Turing machine goes as follows. Start with the head at the left end of the tape, in the *start* state. Then proceed, one step at a time: if the Turing machine has not halted, locate an instruction in S whose condition is satisfied, and execute the action.

Some comments. If the head is at the left end of the tape, and an action calls for a move left, we follow the convention that the head does not move. If at some stage of a computation no instruction has a condition that is satisfied, think of the machine as entering an infinite loop. What is defined above is a *non-deterministic* computation—at some stage of a computation more than one instruction could have a condition that is satisfied. In such a case, arbitrarily choose one applicable instruction, and use that. We could require our Turing machines to be *deterministic*, so that such a choice can never come up. It can be shown that deterministic machines can simulate non-deterministic machines, though runs are longer, so for our purposes placing a deterministic restriction does not matter—we have chosen the non-deterministic version because it is simpler to describe. Note that when analyzing *complexity* of computation, the choice most definitely does matter.

Let us say a Turing machine M is specified by giving a finite set S of states, one of which is designated as a *start* state, and a finite set I of instructions. We say machine M *accepts* a finite word w made up of 0's and 1's provided, if we write word w on the tape starting at the left end, position the Turing machine head on the left end of the tape, and set the state to be *start*, some run of the machine halts. (Note we say *some* run, because we have non-determinism. Not all runs may halt, but we accept the word if at least one does.)

Acceptance is really a *semi*-decision procedure. We accept if the Turing machine halts, and this is something we will eventually discover if we run the machine through enough steps. We don't accept if the machine doesn't halt, but no finite number of steps of a run is ever enough to demonstrate non-halting.

We can also think of a Turing machine as accepting *numbers*. A machine accepts the number n if it halts when started with a base 2 representation for n on the tape.

Here is an example. There are three states $\{start, z, n\}$. (Think of z as "looking for zeros" and n as "no good".) The instructions are as follows.

$$
\begin{array}{rcl}
0, start & \to & 0, n, R \\
1, start & \to & 1, z, R \\
0, z & \to & 0, z, R \\
b, z & \to & halt \\
1, z & \to & 1, n, R \\
0, n & \to & 0, n, R \\
1, n & \to & 1, n, R \\
b, n & \to & b, n, R
\end{array}
$$

Here are two examples using this program. We show a tape by displaying the symbols on it, from the left, omitting blank characters. We underline the position the head scans, and we give the state explicitly. The first example shows that the number 4 (in binary notation, 100) is accepted.

$$
\begin{array}{ll}
\underline{1}00bb\ldots & start \\
1\underline{0}0bb\ldots & z \\
10\underline{0}bb\ldots & z \\
100\underline{b}b\ldots & z \\
halt &
\end{array}
$$

The second example shows 5 (in binary 101) is not accepted—the machine runs forever.

$$
\begin{array}{ll}
\underline{1}01bb\ldots & start \\
1\underline{0}1bb\ldots & z \\
10\underline{1}bb\ldots & z \\
101\underline{b}b\ldots & n \\
101b\underline{b}\ldots & n \\
\vdots & \vdots
\end{array}
$$

Try some examples yourself—this machine accepts numbers that are powers of 2. Notice that this machine does not change symbols—briefly, it does not write.

Basic Fact For each Turing machine, the set of numbers it accepts is Σ.

Actually, it is little more work to establish a stronger result, which is stated below. Suppose we represent each Turing machine instruction as a *set* in some way. We could do this as follows. Take the states $\{s_0, s_1, \ldots, s_n\}$ to be non-negative integers $\{0, 1, \ldots, n\}$. Also think of L and R as 0 and 1. Then think of an instruction like $0, s_4 \to 1, s_5, R$ as $\langle 0, 4, 1, 5, 1 \rangle$. In this way a Turing machine program is just a set.

Even More Basic Fact The following two-place relation is Σ: n is in the set accepted by Turing machine with program S.

Exercises

Exercise 5.1. Prove the "Even More Basic Fact" stated above. (long)

Exercise 5.2. Give instructions for a Turing machine to duplicate a string. More precisely, the machine should start with a blank followed by a string of 0's and 1's, call it S, and it should end with a blank, followed by S, followed by a blank, followed by another copy of S.

Exercise 5.3. Represent the number n by $011\ldots 1$, instead of in binary notation, where this is a sequence of n occurrences of the symbol 1 following the symbol 0. Define a Turing machine that computes the 'doubling' function. More precisely, if the machine starts with a representation of n on its tape, then it halts with a representation of $2 * n$ on its tape.

Exercise 5.4. Give instructions for a Turing machine that computes the successor function in binary. That is, the alphabet is $\{0, 1\}$ and, when the machine is started with a blank followed by a base 2 name on its tape, it returns a blank followed by the base 2 name for the next number. Assume a base 2 name does not begin with 0, except for the name 0 itself.

CHAPTER 7

AXIOMATICS

1 Introduction

This book is about mathematical truth, formal proofs, and the extent to which the first notion is captured by the second. But so far we have discussed only truth—truth in the standard structures for arithmetic and finite set theory. It is time to introduce the notion of provability. But even here we find it useful to begin with semantics. Tarski defined an intuitively satisfying, and now standard, notion of logical consequence. This is a semantic notion, and involves models for its very definition. However, it can be shown that this semantical idea is the exact counterpart of a purely syntactic construction, that of a proof from a set of axioms. This is Gödel's completeness theorem. So, our fundamental question becomes: can we find a 'suitable' set of axioms from which we can derive exactly the truths of arithmetic, or of set theory? This fundamental question will be answered in subsequent chapters. Here we lay the groundwork by introducing the notions of *proof*, of *theory*, and of *formal theory*.

2 Truth in Models

In Section 4 of Chapter 2 we said what it meant for a sentence to be true in a *canonical* model. The use of canonical models simplifies things somewhat. The quantifiers are the central problem. We should take $(\forall x)\varphi(x)$ to be true in a model provided $\varphi(x)$ is true of each thing in the domain of the model. If each thing in the domain has a name, we can simply say $(\forall x)\varphi(x)$ is true if $\varphi(t)$ is true for each name, that is, for each closed term t. But if some item, say m, in the model does not have a name in our language, how can we say $\varphi(x)$ is true if x has the value m? To deal with this we need an additional piece of machinery, that of *assignment*, which is not needed in dealing with canonical models. (The term *valuation* is also used.)

Definition 2.1. Let $L(\mathbf{R}, \mathbf{F}, \mathbf{C})$ be a first-order language, and let $\mathcal{M} = \langle \mathcal{D}, \mathcal{I} \rangle$ be a model for this language (See Definition 3.1, Chapter 2). An *assignment* in this model is a mapping $\mathcal{A}$ from the set of variables of the language to the domain $\mathcal{D}$ of the model. We write $x^{\mathcal{A}}$ for the image of the variable x under the assignment $\mathcal{A}$.

If we have both a model and an assignment we can give a meaning to any term, even if it contains variables. For variables, we use the assignment, and otherwise we proceed as we did in Definition 3.2 of Chapter 2.

Definition 2.2. Let $L(\mathbf{R}, \mathbf{F}, \mathbf{C})$ be a language, $\mathcal{M} = \langle \mathcal{D}, \mathcal{I} \rangle$ be a model for the language, and $\mathcal{A}$ be an assignment in this model. To each term t, closed or not, is assigned a 'meaning,' $t^{\mathcal{M}, \mathcal{A}} \in \mathcal{D}$, as follows.

1. For a constant symbol c, set $c^{\mathcal{M}, \mathcal{A}} = c^{\mathcal{I}}$.

2. For a variable x, set $x^{\mathcal{M}, \mathcal{A}} = x^{\mathcal{A}}$.

3. For an n-place function symbol f, and terms $t_1, \ldots, t_n$, set $[f(t_1, \ldots, t_n)]^{\mathcal{M}, \mathcal{A}} = f^{\mathcal{I}}(t_1^{\mathcal{M}, \mathcal{A}}, \ldots, t_n^{\mathcal{M}, \mathcal{A}})$.

Now it is easy to give a truth value to any formula, closed or not, in any model, canonical or not, but relative to an assignment of values to variables. The following modifies Definition 4.1, Chapter 2.

Definition 2.3. Let $L(\mathbf{R}, \mathbf{F}, \mathbf{C})$ be a language, $\mathcal{M} = \langle \mathcal{D}, \mathcal{I} \rangle$ be a model for the language, and $\mathcal{A}$ be an assignment in this model. The atomic formula (not necessarily a sentence) $P(t_1, \ldots, t_n)$ is true in $\mathcal{M}$ relative to the assignment $\mathcal{A}$ if the n-tuple $\langle t_1^{\mathcal{M}, \mathcal{A}}, \ldots, t_n^{\mathcal{M}, \mathcal{A}} \rangle$ is in the relation $P^{\mathcal{I}}$.

This can be extended to non-atomic formulas, after a simple preliminary definition.

Definition 2.4. Let x be a variable. Two assignments are called x-*variants* provided they assign the same values to every variable except possibly x.

The following modifies Definition 4.2, Chapter 2.

Definition 2.5. Again let $L(\mathbf{R}, \mathbf{F}, \mathbf{C})$ be a language, $\mathcal{M} = \langle \mathcal{D}, \mathcal{I} \rangle$ be a model for the language, and $\mathcal{A}$ be an assignment in this model. Truth for arbitrary formulas of the language L, in the model $\mathcal{M}$, relative to the assignment $\mathcal{A}$, is defined recursively, as follows.

1. Atomic formulas are covered by Definition 2.3.

2. $(A \wedge B)$ is true in $\mathcal{M}$ relative to $\mathcal{A}$ if both A and B are.

3. $(A \vee B)$ is true in $\mathcal{M}$ relative to $\mathcal{A}$ if one of A or B is.

4. $(A \supset B)$ is true in $\mathcal{M}$ relative to $\mathcal{A}$ if A is not or B is.

5. $\neg A$ is true in $\mathcal{M}$ relative to $\mathcal{A}$ if A is not.

6. $(\forall x)\varphi(x)$ is true in $\mathcal{M}$ relative to $\mathcal{A}$ provided $\varphi(x)$ is true in $\mathcal{M}$ relative to each assignment $\mathcal{B}$ that is an x-variant of $\mathcal{A}$.

7. $(\exists x)\varphi(x)$ is true in $\mathcal{M}$ relative to $\mathcal{A}$ provided $\varphi(x)$ is true in $\mathcal{M}$ relative to some assignment $\mathcal{B}$ that is an x-variant of $\mathcal{A}$.

It is not hard to show that if two assignments $\mathcal{A}$ and $\mathcal{B}$ agree on all the free variables of a formula φ, then in a model $\mathcal{M}$ the truth value of φ will be the same relative to $\mathcal{A}$ and to $\mathcal{B}$. Consequently for sentences, truth values are entirely independent of the choice of assignment.

Definition 2.6. A sentence φ is *true* in a model $\mathcal{M}$ provided φ is true in $\mathcal{M}$ relative to some (equally well, to any) assignment $\mathcal{A}$.

With this out of the way, we can introduce the primary notion we have been after all along, that of a sentence being a logical consequence of a set of sentences, or axioms. In what follows, it is understood that a single fixed language is always involved.

Definition 2.7. Let S be a set of sentences, and let φ be a single sentence. We say φ is a *logical consequence* of the set S, and we write $S \models \varphi$ provided φ is true in any model in which all members of S are true.

Exercises

Exercise 2.1. A sentence is called *valid* if it is a logical consequence of the empty set of sentences. Equivalently, a sentence is valid if it is true in all models (for the language of the sentence). Which of the following are valid, and which are not?

1. $(\forall x)(P(x) \supset Q(x)) \supset ((\forall x)P(x) \supset (\forall x)Q(x))$
2. $(\forall x)(P(x) \vee Q(x)) \supset ((\forall x)P(x) \vee (\forall x)Q(x))$
3. $(\exists x)(\forall y)R(x,y) \supset (\forall y)(\exists x)R(x,y)$
4. $(\forall x)(\exists y)R(x,y) \supset (\exists y)(\forall x)R(x,y)$

Exercise 2.2. If $\mathcal{M}$ is a canonical model, we actually have two definitions of truth for sentences. Show they agree.

3 Formal Proofs

The notion of logical consequence is intuitively satisfying—it seems to be what the notion ought to be. But it is quite non-constructive. The definition talks about all models, and the collection of all models is too big to constitute a set, in the technical sense of set used in formal treatments of set theory. Fortunately, Gödel showed that it is equivalent to another notion with a more constructive flavor to it, that of *proof*. Roughly, a proof of φ from a set S is a finite sequence of sentences, constructed according to simple, syntactic rules, starting with members of S, and ending with φ. What Gödel showed is that, if we use a certain family of natural rules, φ is a logical consequence of S if and only if φ has a proof from S. Since Gödel's initial work, many other families of proof rules have been shown to have this property—a combination of *completeness* and *soundness*. We do not prove Gödel's result here; there are many standard books that contain a proof. But we do quickly sketch a complete and sound system, for purposes of reference later on, and because it allows us to establish some fundamental facts about consequence that we will need.

Before we begin with the main details, there is a preliminary issue to be dealt with. Some proof procedures allow the use of arbitrary formulas, others do not. We have chosen to work with a version that does not; only sentences appear in a proof. In this case, then, extra constant symbols must be added to the language for use only in proofs. This is a well-known phenomenon in informal mathematics. Suppose we have established, in an informal argument, that something has the property P. Then we often say, "let c be a name for something that has P." Of course, c must have no prior commitments—in effect we introduce a new constant symbol at this point. Well, for this purpose we must enlarge languages by the addition of new constant symbols called *parameters*, for use in proofs.

Definition 3.1. Let $L = L(\mathbf{R}, \mathbf{F}, \mathbf{C})$ be a language. By $L^{\mathbf{par}}$ we mean an extension of L containing an infinite set of additional constant symbols, called *parameters*. Thus $L^{\mathbf{par}} = L(\mathbf{R}, \mathbf{F}, \mathbf{C} \cup \mathbf{P})$, where $\mathbf{P}$ is infinite, and new.

Now for our proof system. To keep things simple we take only $\neg$ and $\supset$ as basic, treating the other connectives as defined. Likewise we take only $\forall$ as basic, treating $\exists$ as defined. Our rules come in two forms. One type says certain sentences are designated as *logical axioms*. The other type says certain sentences follow from others, *rules of inference*. We begin with the first type.

Propositional Axiom Schemas All sentences of $L^{\mathbf{par}}$ of the following forms are axioms:

1. $(X \supset (Y \supset X))$

2. $((X \supset (Y \supset Z)) \supset ((X \supset Y) \supset (X \supset Z)))$

3. $((\neg Y \supset \neg X) \supset ((\neg Y \supset X) \supset Y))$

First-Order Axiom Schema If t is a closed term of L^{par}, all sentences of L^{par} of the following form are axioms:

$(\forall x)\varphi(x) \supset \varphi(t)$

Items of the forms above are the only logical axioms. Now for the two rules of inference.

Modus Ponens The sentence Y follows from the sentences X and $(X \supset Y)$, or schematically:

$$\frac{X \quad (X \supset Y)}{Y}$$

Universal Generalization The sentence $(\Phi \supset (\forall x)\varphi(x))$ follows from the sentence $(\Phi \supset \varphi(c))$, provided c is a parameter that does not occur in Φ or in $\varphi(x)$. Schematically,

$$\frac{(\Phi \supset \varphi(c))}{(\Phi \supset (\forall x)\varphi(x))}$$

provided c is a parameter not occurring in the conclusion.

This completes the preliminaries. Now the main event.

Definition 3.2. Let S be a set of sentences of L. A *proof* from S is a finite sequence of sentences of L^{par}, each of which is either a member of S, a logical axiom, or follows from earlier lines by one of the rules of inference. A sentence X of L is *provable* from S if there is a proof from S with X as its last line.

Note that sentences of L are proved, but sentences of L^{par} may be used in the proof. In many other books you will easily find examples of proofs using this, or similar, axiom systems. In addition, you will find the following fundamental result, which we use but do not prove here.

Theorem 3.3 (Gödel's Completeness Theorem).
$S \models X$ if and only if X is provable from S.

4 Equality

Let us assume we are working with a language containing a symbol $\approx$ that is intended to represent equality. The question is, what do we need to assume about this relation symbol to be sure it really will represent the equality relation. The solution to this is well-known and, once again, details can be found in standard books. We present only what we will directly need.

The essential properties of equality are that it is an equivalence relation, and one can substitute equals for equals. It is not hard to show that, in the presence of substitutivity, only the reflexive property of equivalence relations needs to be assumed; transitivity and symmetry follow. Consequently, we are led to the following essential properties.

Reflexivity Condition $(\forall x)(x \approx x)$

Substitutivity Conditions

1. If f is an n-place function symbol of L,
   $(\forall x_1)\ldots(\forall x_n)(\forall y_1)\ldots(\forall y_n)\{[(x_1 \approx y_1) \wedge \ldots \wedge (x_n \approx y_n)] \supset f(x_1,\ldots,x_n) \approx f(y_1,\ldots,y_n)\}.$

2. If P is an n-place relation symbol of L,
   $(\forall x_1)\ldots(\forall x_n)(\forall y_1)\ldots(\forall y_n)\{[(x_1 \approx y_1) \wedge \ldots \wedge (x_n \approx y_n)] \supset [P(x_1,\ldots,x_n) \supset P(y_1,\ldots,y_n)]\}.$

The Substitutivity Conditions are stated only for function symbols and relation symbols, but it can be shown that similar conditions follow for arbitrary terms and arbitrary formulas. That is, the following can be shown.

Proposition 4.1. *Let L be a language with a binary relation symbol $\approx$, and let $\mathcal{E}$ be the collection of formula consisting of the Reflexivity Condition and the Substitutivity Conditions for L. Then:*

1. *If $t(x_1,\ldots,x_n)$ is any term of L with x_1, $\ldots$, x_n free,*
   $\mathcal{E} \models (\forall x_1)\ldots(\forall x_n)(\forall y_1)\ldots(\forall y_n)\{[(x_1 \approx y_1) \wedge \ldots \wedge (x_n \approx y_n)] \supset t(x_1,\ldots,x_n) \approx t(y_1,\ldots,y_n)\}.$

2. *If $\varphi(x_1,\ldots,x_n)$ is any formula of L with x_1, $\ldots$, x_n free,*
   $\mathcal{E} \models (\forall x_1)\ldots(\forall x_n)(\forall y_1)\ldots(\forall y_n)\{[(x_1 \approx y_1) \wedge \ldots \wedge (x_n \approx y_n)] \supset [\varphi(x_1,\ldots,x_n) \supset \varphi(y_1,\ldots,y_n)]\}.$

Definition 4.2. A model $\mathcal{M}$ for a language L with an equality symbol $\approx$ is called *normal* if $\approx$ is interpreted in $\mathcal{M}$ as the equality relation on the domain of the model. We say a set S of sentences *satisfies the equality conditions* if the Reflexivity Condition and the Substitutivity Conditions are logical consequences of (equivalently, are provable from) S.

Gödel's Completeness Theorem has the following extension, also due to Gödel.

Theorem 4.3. *Let L be a language with an equality symbol $\approx$, and let X be a sentence and S be a set of sentences of L. If S satisfies the equality conditions, then the following are equivalent:*

1. $S \models X$,

2. X is true in every normal model in which the members of S are true.

Thus, satisfying the equality conditions is exactly what is needed to make sure a relation symbol $\approx$ that is intended to represent equality really does so.

The discussion so far applies to LA, which has a symbol $\approx$ intended to represent equality. What about LS, which doesn't? Typically, in set theory, sets are considered equal if they have the same extension. So, suppose we say $x \approx y$ abbreviates the formula $(\forall z \, \varepsilon \, x)(z \, \varepsilon \, y) \wedge (\forall z \, \varepsilon \, y)(z \, \varepsilon \, x)$, and require that this meets appropriate conditions. That is, we extend Definition 4.2 above.

Definition 4.4. Let L be a language, and suppose there is a designated formula of L with two free variables, which we abbreviate $(x \approx y)$, and which we intend to represent equality. A model $\mathcal{M}$ for this language is *normal* provided that $(x \approx y)$ is true in $\mathcal{M}$ under an assignment $\mathcal{A}$ if and only if $\mathcal{A}(x) = \mathcal{A}(y)$. A set S of sentences of L satisfies the equality conditions if the Reflexivity Condition and the Substitutivity Condition are consequences of S, understanding occurrences of $(x \approx y)$ to be occurrences of the designated formula.

Under this extended definition, Theorem 4.2 continues to hold. This allows us to treat LS, with its defined equality, as well as LA, where equality is primitive.

5 Theories

It is convenient to give the collection of logical consequences of a set of sentences a name; *theory* is standard for this purpose.

Definition 5.1. Let $\mathcal{A}$ be a set of sentences of L. The set of logical consequences of $\mathcal{A}$ in the language L is called the *theory of $\mathcal{A}$*, and is denoted *Theory*($\mathcal{A}$). A set $\mathcal{B}$ is a *theory* if $\mathcal{B} = $ *Theory*($\mathcal{A}$) for some set $\mathcal{A}$.

Then by definition, and Gödel's Completeness Theorem, the following are equivalent:

1. $\mathcal{A} \models X$

2. $X \in $ *Theory*($\mathcal{A}$)

3. X is provable from $\mathcal{A}$

We often refer to the members of $\mathcal{A}$ as *axioms* for *Theory*($\mathcal{A}$). Note that the set of axioms for a theory is not unique, that is, *Theory*($\mathcal{A}$) = *Theory*($\mathcal{B}$) does not imply $\mathcal{A} = \mathcal{B}$.

The following summarizes the basic facts about theories and their axiomatizations. Its proof is left as an exercise.

Proposition 5.2. *Let $\mathcal{A}$ and $\mathcal{B}$ be sets of sentences. Then:*

1. $\mathcal{A} \subseteq$ *Theory*($\mathcal{A}$)

2. $\mathcal{A} \subseteq$ *Theory*($\mathcal{B}$) *implies Theory*($\mathcal{A}$) $\subseteq$ *Theory*($\mathcal{B}$)

3. $\mathcal{A} \subseteq \mathcal{B}$ *implies Theory*($\mathcal{A}$) $\subseteq$ *Theory*($\mathcal{B}$)

4. *Theory*(*Theory*($\mathcal{A}$)) = *Theory*($\mathcal{A}$)

Notice that by the last item of the Proposition above, *Theory*($\mathcal{A}$) itself will serve as a set of axioms for *Theory*($\mathcal{A}$). We also have the following, which serves as a rich and usually uninteresting source of theories.

Proposition 5.3. *Let $\mathcal{M}$ be a model for the language L, and let $\mathcal{A}$ be the set of sentences of L that are true in $\mathcal{M}$. Then Theory*($\mathcal{A}$) = $\mathcal{A}$, *so $\mathcal{A}$ is a theory.*

Also, we will generally need a notion of equality available. We do the obvious thing.

Definition 5.4. A theory $\mathcal{T}$ is a theory *with equality* if the equality conditions of Section 4 are members, where $(x \approx y)$ is either atomic, as in LA, or defined, as in LS.

Notation Convention From now on, when we use the term *theory* we assume it is a theory with equality.

There are a few more pieces of terminology that will be useful as we go along. First, we are primarily interested in theories that capture the behavior of $\mathbb{HF}$ or $\mathbb{N}$, so we are interested in theories that are, in some sense, true. The following makes this precise.

Definition 5.5. A theory $\mathcal{T}$ in the language LS is sometimes called a *true theory* if all its members are true in the standard model $\mathbb{HF}$, that is, if $\mathcal{T} \subseteq \mathcal{TS}$. Likewise if $\mathcal{T}$ is in the language LA it is a true theory, or sound, if its members are true in the standard model for arithmetic, $\mathbb{N}$; equivalently, if $\mathcal{T} \subseteq \mathcal{TA}$.

If theory $\mathcal{T}$ has axiomatization $\mathcal{A}$, $\mathcal{T}$ will be a true theory if all members of $\mathcal{A}$ are true in the appropriate standard model. Note that the very concept of true theory is not a constructive one since verifying behavior in infinite models is involved. Nonetheless, it is a useful one.

We will sometimes need to know if a theory is 'sufficiently strong.' The next item embodies a minimal requirement on theories, but one that is strong enough for many of our purposes.

Definition 5.6. A theory T in the language LS is called Δ_0 *complete* if all closed instances of Δ_0 formulas that are true in $\mathbb{HF}$ are in T. Likewise a theory T in the language LA is Δ_0 *complete* if all closed instances of the arithmetic version of Δ_0 formulas that are true in $\mathbb{N}$ are in T.

From now on an assumption of Δ_0 completeness will generally be made. On the one hand it is a weak assumption—it amounts to saying the theory can prove all facts whose truth we can check in the most concrete of ways. On the other hand, Δ_0 completeness endows a theory with a certain strength. The following result, while simple, begins to give some idea of this strength.

Proposition 5.7. *Let T be a theory, either in the language of set theory or of arithmetic.*

1. *If T is Δ_0 complete, then all true closed instances of Σ_1 formulas are in T.*

2. *If T is both Δ_0 complete and a true theory, then a closed instance of a Σ_1 formula is true if and only if it is in T.*

Proof For part 1, suppose $(\exists x)\varphi(x)$ is a closed instance of a Σ_1 formula in the language LS, and it is true in $\mathbb{HF}$. Then for some closed term t, $\varphi(t)$ is true in $\mathbb{HF}$. But this is a true closed instance of a Δ_0 formula so, since T is Δ_0 complete, it is in T. Theories are closed under logical consequence, and $\varphi(t) \supset (\exists x)\varphi(x)$ is a validity, hence $(\exists x)\varphi(x)$ must be in T. The arithmetic version is similar. Part 2 is trivial. ∎

Now that the notion of a theory has been introduced, it makes it possible to ask whether arithmetic or finite set theory has a 'reasonable' axiomatization. But theories are still much too general—after all, by Proposition 5.3 the set of sentences $\mathcal{S}$ in the language of set theory that are true in the standard model $\mathbb{HF}$ constitutes a theory (with equality), and by Proposition 5.2 part 4, $\mathcal{S}$ itself will serve as an axiomatization. Similarly for arithmetic, of course. Surely this is not the sort of thing we had in mind; it simply says arithmetic truth is axiomatized by the set of arithmetic truths, and similarly for hereditarily finite set theory. While correct, this is less than entirely satisfying.

The problem comes down to characterizing what 'reasonable' should mean for an axiom set. And here there is one basic condition we can impose that is uncontroversial: we should be able to tell what is an axiom and what is not. That is, there should be a decision procedure for the set of axioms. In Chapter 5 we showed how formulas of LS, the language of set theory, could be thought of as members of $\mathbb{HF}$. A similar thing can be done with any first-order language, and we assume it has been done for what follows. Then a first attempt at saying what is a 'reasonable' axiomatization for a theory is: the set of axioms should be Δ, or equivalently, the set of Gödel numbers of axioms shoud be recursive.

Many years ago W. Craig proved that decidability was a condition on axiom sets that could be relaxed. We do not give the proof here, but we do state his result.

Theorem 5.8 (Craig's Theorem). *For each Σ set $\mathcal{A}$ of sentences there is a Δ set $\mathcal{B}$ of sentences such that $Theory(\mathcal{A}) = Theory(\mathcal{B})$.*

Loosely, a theory with a semi-decidable set of axioms also has a decidable axiomatization. Craig actually showed his result in an arithmetic framework. Then instead of working with formulas thought of as sets, one works with Gödel numbers of formulas, and instead of Σ and Δ one talks about recursive enumerability and recursiveness. But this is just a variation on what we said.

Given Craig's Theorem, we can relax our 'reasonableness' condition on axioms, leading to our final concept.

Definition 5.9. A *formal theory* is a theory that has a Σ axiomatization. That is, $\mathcal{S}$ is a formal theory if $\mathcal{S} = Theory(\mathcal{A})$ for some Σ set $\mathcal{A}$.

The set $\mathcal{TS}$ of true sentences of $\mathbb{HF}$ is a theory, as we observed above, with itself as one axiomatization. But by Tarski's Theorem (Theorem 8.2, Chapter 5) $\mathcal{TS}$ is not representable, so in particular it is not Σ. Similar comments apply to $\mathcal{TA}$, the set of true sentences of $\mathbb{N}$. We can now raise a very fundamental question. Both $\mathcal{TA}$ and $\mathcal{TS}$ are theories. Are they *formal* theories? The fact that $\mathcal{TS}$ is not Σ does not tell us whether or not there is some other axiomatization for $\mathcal{TS}$ that is Σ. Do $\mathcal{TS}$ and $\mathcal{TA}$ have reasonable axiomatizations? This will be answered at the end of this Chapter, and in the next as well.

Exercises

Exercise 5.1. Prove Proposition 5.2. Show the first two parts directly, and show the last two follow from the first two.

Exercise 5.2. Prove Proposition 5.3.

6 Examples of Formal Theories

It is time to give some concrete examples. We begin with arithmetic, then move on to set theory. Not all the formal theories are intended to be complete—some have been introduced for particular technical purposes, but all the theories here are true theories, in the sense of Definition 5.5.

6.1 Formal Arithmetic Theories

We begin with the most prominent example, Peano Arithmetic. This is an extremely powerful theory, in which one can prove all the basic facts one might think of about arithmetic. There was hope, until Gödel, that it was an axiomatization of $\mathcal{TA}$, the set of sentences true in $\mathbb{N}$ but it is not, as we will see shortly. Nonetheless it is of considerable importance for its own sake, and has been extensively investigated.

The language for Peano Arithmetic is LA. There are several straightforward axioms, and a single axiom schema. We begin with the axioms, starting with those for equality, which makes it a theory with equality.

Peano Axioms

Equality Axioms:

1. $(\forall x)(x \approx x)$
2. $(\forall x_1)(\forall x_2)(\forall y_1)(\forall y_2)\{[(x_1 \approx y_1) \wedge (x_2 \approx y_2)] \supset [(x_1 \oplus x_2) \approx (y_1 \oplus y_2)]\}$
3. $(\forall x_1)(\forall x_2)(\forall y_1)(\forall y_2)\{[(x_1 \approx y_1) \wedge (x_2 \approx y_2)] \supset [(x_1 \otimes x_2) \approx (y_1 \otimes y_2)]\}$
4. $(\forall x_1)(\forall y_1)\{(x_1 \approx y_1) \supset (\mathbb{S}(x_1) \approx \mathbb{S}(y_1))\}$
5. $(\forall x_1)(\forall x_2)(\forall y_1)(\forall y_2)\{[(x_1 \approx y_1) \wedge (x_2 \approx y_2)] \supset [(x_1 \approx x_2) \supset (y_1 \approx y_2)]\}$

Arithmetic Axioms:

1. $(\forall x)\neg(\mathbb{S}(x) \approx \mathbf{0})$
2. $(\forall x)(\forall y)[(\mathbb{S}(x) \approx \mathbb{S}(y)) \supset (x \approx y)]$
3. $(\forall x)[(x \oplus \mathbf{0}) \approx x]$
4. $(\forall x)(\forall y)[(x \oplus \mathbb{S}(y)) \approx \mathbb{S}(x \oplus y)]$
5. $(\forall x)[(x \otimes \mathbf{0}) \approx \mathbf{0}]$
6. $(\forall x)(\forall y)[(x \otimes \mathbb{S}(y)) \approx ((x \otimes y) \oplus x)]$

Finally, there is one axiom schema, that for induction.

Induction Schema All sentences of the following form are axioms:

$$\{\varphi(\mathbf{0}) \wedge (\forall x)[\varphi(x) \supset \varphi(\mathbb{S}(x))]\} \supset (\forall x)\varphi(x)$$

Let PA, for Peano Arithmetic, be the set of axioms given above. It is easy to show that this is a Σ set (Δ even), and so *Theory*(PA) is a formal theory. As we remarked above, it is a formal theory of considerable strength. Unfortunately, as we will see, *Theory*(PA) $\neq \mathcal{TA}$.

Another formal arithmetic theory (with equality) of interest is the theory Q, due to R. Robinson. It was never intended to be an axiomatization of $\mathcal{TA}$, but rather to have particular theoretical applications. Peano arithmetic has infinitely many axioms, because of the Induction Schema, and it can be shown that *Theory*(PA) has no finite axiomatization. Robinson's system Q, though weaker, does have a finite axiomatization, and this is why it was created. Its axioms are those of PA, except for the induction schema, but with the following single axiom added:

$$(\forall x)\{(x \approx \mathbf{0}) \vee (\exists y)(x \approx \mathbb{S}(y))\}$$

6.2 Formal Set Theories

Next we turn to some systems for set theory. The intended model now is $\mathbb{HF}$, and this is like the entire universe of sets, except that everything is finite. So, a natural approach is to use the axioms of Zermelo-Fraenkel set theory, but with the axiom of infinity replaced by its negation. This gives a theory that is inter-translateable with PA, using essentially the same translation we gave between $\mathbb{HF}$ and $\mathbb{N}$. We refer to this axiomatization as ZF $- \infty$. Its language is not *LS*, but instead can be formulated with only ε as primitive, or with this and also a constant symbol for the empty set and function symbols for unordered pair, union, and power set.

Another axiomatization that is more natural in our present setting involves making use of the function and constant symbols of the language *LS*. This means we give an axiom for $\mathbb{A}$ instead of axioms for unions, pairs, and power sets. This gives rise to the following system, which we call FIN_SET. In stating the axioms, for convenience we use $(x \approx y)$ to abbreviate $(\forall q \, \varepsilon \, x)(q \, \varepsilon \, y) \wedge (\forall q \, \varepsilon \, y)(q \, \varepsilon \, x)$.

FIN_SET Axioms, Part 1

1. (Extensionality) $(\forall x)(\forall y)[x \approx y \supset (\forall z)(x \, \varepsilon \, z \supset y \, \varepsilon \, z)]$

2. (Empty Set) $(\forall x)\neg(x \, \varepsilon \, \emptyset)$

3. (Addition) $(\forall x)(\forall y)(\forall z)[z \, \varepsilon \, \mathbb{A}(x,y) \equiv (z \, \varepsilon \, x \vee z \approx y)]$

4. (Foundation) $(\forall x)\{(\exists y)(y \, \varepsilon \, x) \supset (\exists y)[y \, \varepsilon \, x \wedge \neg(\exists z)(z \, \varepsilon \, x \wedge z \, \varepsilon \, y)]\}$

You showed the Foundation Axiom is true in $\mathbb{HF}$, in Exercise 4.4. The next item is an axiom scheme, and so represents an infinite collection of axioms.

FIN_SET Axioms, Part 2 (Replacement) The following is an axiom for each formula $\varphi(x, y, w_1, \ldots, w_n)$, where all free variables are displayed.

$$(\forall w_1)\ldots(\forall w_n)(\forall a)\bigg\{(\forall x)\Big[x \, \varepsilon \, a \supset (\exists y)\big[\varphi(x,y,w_1,\ldots,w_n)$$
$$\wedge (\forall z)\,(\varphi(x,z,w_1,\ldots,w_n) \supset z \approx y)\big]\Big]$$
$$\supset (\exists b)(\forall x)\Big[x \, \varepsilon \, a \supset$$
$$(\exists y)(y \, \varepsilon \, b \wedge \varphi(x,y,w_1,\ldots,w_n))\Big]\bigg\}$$

Interpreted in $\mathbb{HF}$, the Replacement Scheme roughly says that any subset of R_ω that is smaller than a member, is a member. The antecedent says φ defines a function with domain a. The consequent says there is a member of R_ω, b, that is range of this function. Finally we have the negation of the axiom of infinity.

FIN_SET Axioms, Part 3 (Finiteness)

$$\neg(\exists x)[\emptyset \, \varepsilon \, x \wedge (\forall z)(z \, \varepsilon \, x \supset \mathbb{A}(z,z) \, \varepsilon \, x)]$$

Notice that the Finiteness axiom just says the collection of numbers does not exist (recall that while $\omega \subseteq R_\omega$, we also have $\omega \notin R_\omega$). One can use the Replacement axioms to establish the non-existence of other infinite sets.

Because of the Replacement axiom scheme, the axiom set for **FIN_SET** is infinite. Here is an analog of Robinson's theory Q for arithmetic. It is important to note that it involves only a *finite* set of axioms. We call it **SET_Q**. The axioms are simply those numbered 1–3 in Part 1 of the **FIN_SET** axiomatization. **SET_Q** is a weak theory, but even so it is Δ_0 complete, Definition 5.6. You are asked to show this in Exercises 6.3-6.6. This will have an important consequence in Chapter 9.

6.3 Theories You Might Not Have Expected

We conclude the chapter with a few formal theories that are somewhat less conventional. Let $\mathcal{S}_\Sigma$ be the set of instances of Σ formulas of LS that are true in the standard model $\mathbb{HF}$. According to Theorem 3.5, Chapter 6, this is a Σ set, and so $Theory(\mathcal{S}_\Sigma)$ is a formal theory. It is even a theory with equality. This is quite different from the examples above, in that the axioms don't fall into a small number of easily recognizable patterns, but nonetheless the conditions for being a formal theory are met. We are being overly generous, perhaps, in what we accept as an axiomatization, but then if we show something is *not* a formal theory, we have shown something quite strong. Exercise 6.2 asks you to show that $Theory(\mathcal{S}_\Sigma)$ is Δ_0 complete.

Naturally, there is a similar example arising from arithmetic. Let $\mathcal{A}_\Sigma$ be the set of closed instances of Σ formulas of LA that are true in $\mathbb{N}$. By Exercise 3.3 of Chapter 6 this is Σ, and so $Theory(\mathcal{A}_\Sigma)$ is also a formal theory (again, with equality).

Exercises

Exercise 6.1. Show FIN_SET is a theory with equality.

Exercise 6.2. Prove that $Theory(\mathcal{S}_\Sigma)$ is Δ_0 complete.

Exercise 6.3. Let $t_1, t_2, \ldots$ be closed terms of LS. And let us write $\langle t_1 \rangle$ for the LS term $\mathbb{A}(\varnothing, t_1)$, and $\langle t_1, t_2, \ldots, t_n, t_{n+1} \rangle$ for $\mathbb{A}(\langle t_1, \ldots, t_n \rangle, t_{n+1})$. Show that for each n the following is provable from the axioms SET_Q.

$$(\forall x)[x \, \varepsilon \, \langle t_1, \ldots, t_n \rangle \supset (x \approx t_1 \vee \ldots \vee x \approx t_n)]$$

Exercise 6.4. We use the R_n sequence from Definition 4.2, Chapter 1. Call n *positively good* provided: 1) for each $s_1, s_2 \in R_n$, and for any closed LS terms t_1 naming s_1 and t_2 naming s_2, if $s_1 \in s_2$ then $t_1 \, \varepsilon \, t_2$ is provable from the axioms SET_Q; 2) if $s \in R_n$ and both t_1 and t_2 name s then $t_1 \approx t_2$ is provable from SET_Q. Show that each n is positively good.

This can be done by induction. The key item needed is that any member of an hereditarily finite set s must be a set of lower rank than s. Here is an outline of the proof. 0 is trivially positively good. Assume n is positively good.

1. Suppose $s_1, s_2 \in R_{n+1}$ and $s_1 \in s_2$. Let t_1 name s_1 and t_2 name s_2. If s_1 and s_2 are, in fact, in R_n we are done. s_1 cannot be in R_{n+1} but not in R_n. So it is enough to show the result under the assumption that $s_1 \in R_n$ and $s_2 \in R_{n+1}$. Now show $t_1 \, \varepsilon \, t_2$ is provable from SET_Q, by an induction on the number of occurrences of $\mathbb{A}$ in t_2.

2. Suppose $s \in R_{n+1}$ and t_1 and t_2 both name s; it must be shown that $t_1 \approx t_2$ is provable from **SET_Q**. Now, make use of Exercise 6.3 and part 1 of this Exercise.

Exercise 6.5. As in the previous exercise, we use terminology from Definition 4.2, Chapter 1, again. Call n *negatively good* provided: 1) for any two distinct $s_1, s_2 \in R_n$, and for any closed LS terms t_1 naming s_1 and t_2 naming s_2, $\neg(t_1 \approx t_2)$ is provable from the axioms **SET_Q**; 2) for any $s_1 \in R_n$ and $s_2 \in R_{n+1}$ with $s_1 \notin s_2$, and for any closed terms t_1 naming s_n and t_2 naming s_2, $\neg(t_1 \, \varepsilon \, t_2)$ is provable from **SET_Q**. Show that each n is negatively good.

To show this use induction, and in the induction step establish part 1 before showing part 2. You will need Exercise 6.3.

Exercise 6.6. Show by induction on degree that, if $\varphi(x_1, \ldots, x_n)$ is any Δ_0 formula, and $t_1, \ldots, t_n$ are closed terms of LS,

1. if $\varphi(t_1, \ldots, t_n)$ is true in $\mathbb{HF}$ then $\varphi(t_1, \ldots, t_n)$ is provable from **SET_Q**;

2. if $\varphi(t_1, \ldots, t_n)$ is false in HF then $\neg \varphi(t_1, \ldots, t_n)$ is provable from **SET_Q**.

CHAPTER 8

GÖDEL'S THEOREM

1 Gödel's Theorem, Tarski's Proof

In the previous chapter we raised a fundamental question about $\mathbb{N}$ and $\mathbb{HF}$: can the notion of truth in these structures be captured axiomatically. Gödel gave an answer to this, with his famous First Incompleteness Theorem. In this section we prove a version of it. We do not, however, give Gödel's original proof. Instead we give one due to Tarski which most people find easier. And besides, we already have almost all the pieces necessary for it. In the next section we discuss the limitations of Tarski's proof, after which we present Gödel's original argument.

There is one item missing before we get to Gödel's theorem. And its proof is of a kind you have seen before, so we leave it as Exercise 1.1.

Proposition 1.1. *Suppose $\mathcal{A}$ is a set of sentences that is Σ in $\mathbb{HF}$. Then so is Theory($\mathcal{A}$).*

Now, here is one version of Gödel's theorem. Recall, $\mathcal{TS}$ is the set of sentences of LS that are true in $\mathbb{HF}$, and $\mathcal{TA}$ is the set of sentences of LA that are true in $\mathbb{N}$.

Theorem 1.2 (Gödel's Theorem, Version 1). *Neither $\mathcal{TS}$ nor $\mathcal{TA}$ is a formal theory.*

Proof If $\mathcal{TS}$ were a formal theory, it would have a Σ axiomatization. Then by Proposition 1.1, $\mathcal{TS}$ itself would be Σ. But by Tarski's Theorem 8.2 of Chapter 5, $\mathcal{TS}$ is not even representable, let alone Σ. So $\mathcal{TS}$ is not a formal theory. The result for $\mathcal{TA}$ follows similarly, using Theorem 10.1, Chapter 5. ∎

This is actually quite devastating. Not only is $\mathcal{TS}$ not a formal theory, but even the most basic of our mathematical structures, $\mathbb{N}$, the one we learned about as children, can not be fully understood using first-order logic and the axiomatic method! Any axiom system that might be proposed cannot prove exactly the truths of arithmetic—either something false must be provable, or something true must be unprovable. If we look at Peano Arithmetic, $\mathcal{PA}$, it is clear that all its axioms are, in fact, true in $\mathbb{N}$. It follows that all the sentences provable from PA are also true in $\mathbb{N}$. Thus we have the following, which is closer in wording to Gödel's theorem as he originally stated it.

Corollary 1.3. *Peano Arithmetic is incomplete; there is some sentence of* LA *that is true in the standard model, but that is not provable from* PA.

There are, of course, similar incompleteness results for both ZF $-\infty$ and FIN_SET. Later on, when we come to Gödel's Second Incompleteness Theorem, we will give a rather remarkable example of a true but unprovable sentence.

Exercises

Exercise 1.1. Prove Proposition 1.1. Hint: replace logical consequence by provability, then use as a model the way we showed the set of terms (S-25) and the set of formulas (S-29) was representable.

Exercise 1.2. Show a theory is a formal theory if and only if it is Σ.

2 Finitary Proofs

In our work so far we have made free use of the notion of truth. This is mathematically well-defined, but it is not, in general, subject to human determination. To check whether $(\forall x)\varphi(x)$ is true in $\mathbb{HF}$, we must check the truth of infinitely many instances. This is not a task for human beings.

It is important to note when a proof uses infinitary concepts, like truth in a model, and when it does not. There are those who, on philosophical grounds, totally reject the use of infinitary notions in mathematics. There are those, a greater number, who feel some uneasiness about their use, especially when it comes to foundational issues. In the 1920's especially, there was much dispute about the foundations of mathematics. On one side were Brouwer and the intuitionists. On the other side were Hilbert and his followers. The intuitionists objected to any use of non-constructive methods. Hilbert had conceived of a remarkable way to 'save' traditional mathematical practice from the attacks of the intuitionists. First, one could represent various branches of mathematics as formal theories, within which non-constructive methods could be allowed. Second, one could prove from the outside that the formal theories did represent their subject matter adequately, and were free of contradiction, but these proofs about the formal theories should be entirely constructive in nature—acceptable to an intuitionist.

Gödel's First and Second Incompleteness Theorems thoroughly demolished Hilbert's program, but their impact would have been somewhat lessened if they themselves had proofs that used non-constructive methods.

Consequently Gödel's original argument was along different lines than the one we presented, which used Tarski's Theorem. And, as is commonly the case, changing the argument in ways that avoids non-constructive notions brings with it useful side effects. In this instance, Gödel's Second Theorem uses not just the statement of the First Theorem, but his proof of it as well.

Let us call a mathematical concept *finitary* if it makes no reference, open or hidden, to infinite sets, that is, to infinite sets as completed things. A proof is *finitary* if it involves only finitary concepts, and is constructive. That is, if it asserts something exists, it provides a means of finding it; if it asserts something can be done, it provides a method for doing it in a finite number of steps.

We have not given a precise definition of finitary, only a general, and quite loose, description. An exact characterization is a matter of some controversy. Nonetheless, even from our vague description it should be possible to recognize certain arguments as being finitary, and certain others as not being so. For our purposes, this is sufficient.

In the rest of this Chapter we will be attempting to replace earlier non-finitary arguments by finitary ones. In fact some of our results were established by finitary means, and others can be rephrased so that they take on constructive content. It is worthwhile to review these points.

In earlier chapters several sets were shown to be Σ. In each case an explicit Σ formula was given. And while truth in $\mathbb{HF}$ is not a finitary notion, truth for instances of Σ formulas is, in the sense that if a Σ formula is true, there is a procedure we can follow that will eventually let us determine that fact.

In the previous Chapter the notion of logical consequence was defined. This is clearly non-constructive. But the notion of proof can substitute for it (the verification that the two are equivalent is, not surprisingly, not constructive). In the previous section it was established that the set of consequences of a Σ set $\mathcal{A}$ of axioms is again a Σ set. If you did Exercise 1.1 you used formal proofs, rather than models, and you actually gave a receipt for turning a Σ formula for $\mathcal{A}$ into a Σ formula for *Theory*($\mathcal{A}$). In this sense Proposition 1.1 has a constructive proof, involving only finitary notions. It is in this spirit that we want to re-do the proof of Theorem 1.2, avoiding mention of truth.

3 Representability in a Theory

The fundamental notion involved in both the proof and the very statement of Tarski's Theorem is that of *representability*. But it uses the concept of truth in a model in an essential way, and it is this kind of thing we want to avoid. The obvious thing to do is replace the notion of truth, which is non-constructive, by provability from a (Σ) set of axioms, which is constructive. We introduce an official version in this section, and explore the consequences in this and subsequent Chapters.

There is one minor annoyance that we must deal with, in order to have things run smoothly. Say we are working with LA and the standard model, $\mathbb{N}$, for arithmetic. The number n is in the set that $\varphi(x)$ represents if $\varphi(t)$ is true in $\mathbb{N}$, where t is a closed term that names the number n. Now, it cannot happen that two terms t and u name the same number, n, but $\varphi(t)$ is true in $\mathbb{N}$ while $\varphi(u)$ is not. The definition of truth in a model makes this impossible. But if we try to replace truth by provability, say from a set $\mathcal{A}$ of axioms, things may not be so nice. It can easily happen that $\mathcal{A} \models \varphi(t)$ but $\mathcal{A} \not\models \varphi(u)$. For example, suppose $\mathcal{A}$ consists of sentences embodying the equality principles of Chapter 7, Section 4, and nothing else. Then if $\varphi(x)$ is the formula $x \approx \mathbf{0}$, we have $\mathcal{A} \models \varphi(\mathbf{0})$ but $\mathcal{A} \not\models \varphi(\mathbf{0} \oplus \mathbf{0})$, even though $\mathbf{0}$ and $\mathbf{0} \oplus \mathbf{0}$ name the same number, 0, because we have no axioms governing the intended behavior of $\oplus$. So, should we take 0 as being in the set $\varphi(x)$ represents, using axioms $\mathcal{A}$, or not? Our solution to this problem is to make sure it does not happen. After all, a set of axioms too weak to prove $(\mathbf{0} \oplus \mathbf{0}) \approx \mathbf{0}$ is not a good candidate for an axiomatization of arithmetic anyway.

Proposition 3.1. *We work with either the language* LS *or the language* LA. *Let* $\mathcal{T}$ *be a theory that is* Δ_0 *complete (Definition 5.6, Chapter 7). If t and u are closed terms that name the same thing in the standard model then* $(t \approx u) \in \mathcal{T}$, *and so if* $\varphi(x)$ *is any formula,* $\varphi(t) \in \mathcal{T}$ *if and only if* $\varphi(u) \in \mathcal{T}$.

Proof If t and u name the same thing in the standard model, $(t \approx u)$ is true in the standard model. If we are working with arithmetic this is an atomic formula; if we are working with sets it abbreviates $(\forall z \, \varepsilon \, t)(z \, \varepsilon \, u) \wedge (\forall z \, \varepsilon \, u)(z \, \varepsilon \, t)$. Either way, it is a true instance of a Δ_0 formula and so, by Δ_0 completeness, it is in $\mathcal{T}$. Since $\mathcal{T}$ is a theory with equality, $\varphi(t)$ and $\varphi(u)$ will be equivalent in it, by Proposition 4.1, Chapter 7. ∎

Every formal theory given in Chapter 7, Section 6 is Δ_0 complete. Now we replace representability, as used in earlier chapters, by representability in a Δ_0 complete theory. We do the work in the context of sets, but a similar thing can be done with arithmetic. Recall from Definition 9.1, Chapter 5, that for a set s, we write $\ulcorner s \urcorner$ for any closed term of LS that names s.

Definition 3.2. *Let* $\mathcal{T}$ *be a theory in the language* LS *that is* Δ_0 *complete. We say* $\varphi(x)$ *represents the set* S *in the theory* $\mathcal{T}$ *if, for* $s \in R_\omega$,

$$s \in S \text{ if and only if } \varphi(\ulcorner s \urcorner) \in \mathcal{T}.$$

We say S is representable in the theory T if it is represented in T by some formula. Representability of a set of numbers using an arithmetically Δ_0 complete theory is defined in the same way.

The set TS, consisting of the sentences of LS true in $\mathbb{HF}$, is obviously a Δ_0 complete theory. It is easy to see that representability in this theory amounts to the same thing as representability, defined earlier (Definition 5.1, Chapter 2). Similarly for TA and arithmetic. Consequently our present notion of representability directly generalizes the earlier one.

It might be questioned whether our generalization of representability really *generalizes* things. The following observations show the answer is yes— there are powerful theories in the language LS for which representability is quite different than representability in TS, and similarly for arithmetic, of course. The set of false closed instances of Σ formulas is certainly representable in $\mathbb{HF}$ and hence in the theory TS (since the set of true instances of Σ formulas is Σ, by Theorem 3.5, Chapter 6, and we have negation available). Theorem 2.2, Chapter 7, says the set of false instances of Σ formulas is not Σ. Thus there is a set that is representable in TS that is not Σ. We already know that the theory TS is not a *formal* theory. The following says that for formal theories things are quite different—only Σ sets are representable.

Proposition 3.3. *For either arithmetic or set theory, if a set S is representable in a formal Δ_0 complete theory, S is Σ.*

Proof We work with LS for convenience. The argument for LA is similar. Let T be a formal, Δ_0 complete theory in which S is representable. By Proposition 1.1, T itself is Σ; say $C_T(v_0)$ represents it in $\mathbb{HF}$. Also, S is representable in T, say by the formula $\varphi(v_0)$. Then

$$s \in S \iff \varphi(\ulcorner s \urcorner) \in T$$
$$\iff C_T(\ulcorner \varphi(\ulcorner s \urcorner) \urcorner) \text{ is true in } \mathbb{HF}$$

Let φ' be a closed term naming the formula $\varphi(v_0)$. Then the following is a Σ formula representing S in $\mathbb{HF}$:

$$(\exists t)\, [\mathsf{Names}(t, s) \wedge (\exists u)\, [u \text{ is } \varphi'(t) \wedge C_T(u)]]$$

∎

Notice that this Proposition was given a constructive proof. Suppose $\varphi(v_0)$ represents a set S in a Δ_0 theory with a Σ set of axioms $\mathcal{A}$. We know *Theory*$(\mathcal{A})$ will also be Σ, and your solution of Exercise 1.1 should have showed how to write out a Σ formula for *Theory*$(\mathcal{A})$, given a Σ formula for

$\mathcal{A}$. Then the proof of the Proposition above explicitly gives a Σ formula for S, using the formula $\varphi(v_0)$ that represents it in $Theory(\mathcal{A})$.

Incidentally, the Proposition above also gives us an alternate proof of Theorem 1.2. There is a non-Σ set representable in the theory $\mathcal{TS}$, but if $\mathcal{A}$ is a formal theory, the only sets representable in it are Σ. Consequently $\mathcal{TS}$ can not be a formal theory.

Finally we have a result that will considerably simplify things later on. If a theory meets certain rather straightforward conditions, then many useful things must be representable in the theory. All the theories of Chapter 7, Section 6 meet the conditions of this Proposition. For several of them, such as PA, this is somewhat tedious to show. But it is almost a triviality for $Theory(\mathcal{S}_\Sigma)$, from Chapter 6, Section 6.3, and likewise for $Theory(\mathcal{A}_\Sigma)$, and you were asked to show it for $Theory(SET\_Q)$.

Proposition 3.4. *For either arithmetic or set theory, if a theory $\mathcal{T}$ is Δ_0 complete and is a true theory, then every Σ relation is representable in $\mathcal{T}$.*

Proof We'll state the proof for sets, though a similar argument works for arithmetic. Suppose S is Σ in $\mathbb{HF}$. By Theorem 6.2, Chapter 3, S is represented by a Σ_1 formula in $\mathbb{HF}$, say by $(\exists x)\varphi(x, v_0)$, where φ is Δ_0. This same formula also represents S in the theory $\mathcal{T}$; this is an immediate consequence of part 2 of Proposition 5.7, Chapter 7. ∎

Exercises

Exercise 3.1. A theory is called *inconsistent* if it contains every sentence.

1. Show: if $\mathcal{A} \models X$ and $\mathcal{A} \models \neg X$ for some sentence X, then $Theory(\mathcal{A})$ is inconsistent, and conversely.

2. Suppose $Theory(\mathcal{A})$ is an inconsistent theory. Is it Δ_0 complete? What sets are representable in it?

4 Ordinary Splits

Before reading this section, go back and re-read the proof of Tarski's Theorem in Section 8 of Chapter 5. What we want to do now is reproduce that proof, as far as possible, but replacing truth with provability in a formal theory. In Chapter 5, Section 8 we called a representing formula *ordinary* if it did not belong to the set it represented. Then if φ is a representing formula, it is ordinary if $\varphi \notin \varphi_S$. Representability meant representability in $\mathbb{HF}$ (or in $\mathbb{N}$). Elaborating, this means φ is ordinary if either of the following two equivalent conditions is met:

1. $\neg\varphi(t)$ is true, where t names $\varphi(v_0)$;

2. $\varphi(t)$ is not true, where t names $\varphi(v_0)$.

Let $\mathcal{A}$ be a set of axioms for a Δ_0 complete theory, and let us try to come up with an analog of being ordinary for *Theory*($\mathcal{A}$); we might call it $\mathcal{A}$-ordinary. Reasonably, we replace truth in $\mathbb{HF}$ by provability from $\mathcal{A}$. But then items 1) and 2) above give us two alternatives that need not be equivalent:

1. $\neg \varphi(t) \in$ *Theory*($\mathcal{A}$), where t names $\varphi(v_0)$;

2. $\varphi(t) \notin$ *Theory*($\mathcal{A}$), where t names $\varphi(v_0)$.

The first says φ is in the set represented by $\neg\varphi$ in *Theory*($\mathcal{A}$). The second says φ is not in the set represented by φ in *Theory*($\mathcal{A}$). If $\mathcal{A}$ is inconsistent, every representing formula, in fact, represents R_ω (See Exercise 3.1), and so item 1) will be the case, but not item 2). On the other hand $\mathcal{A}$ might be incomplete, or too weak to prove every sentence or its negation. Then item 2) may very well happen without item 1) being true. When we were working with representability in $\mathbb{HF}$ we were, in effect, using the theory $\mathcal{TS}$, and this is both consistent and complete, so the two notions were equivalent. For arbitrary theories we have a question: which is the 'right' version of ordinary to use? The odd answer is that both are. Each gives us interesting results when we try to mimic the proof of Tarski's Theorem. We investigate alternative 1) here, and turn to alternative 2) in the next Chapter.

5 Gödel's Theorem, Gödel's Proof (more or less)

In the last section we saw that the notion of ordinary splits when we deal with formal theories. Now we follow up on the consequences of taking the first alternative. Since this leads to a proof of Gödel's Theorem that is rather close to the one Gödel himself gave, we call this *ordinary$_G$*. For convenience we carry out the work entirely in the context of set theory: the language is LS, and the intended model is $\mathbb{HF}$. We leave the simple adaptation to arithmetic to you.

Definition 5.1. Let $\mathcal{T}$ be a Δ_0 complete theory (not necessarily formal), and let $\varphi(v_0)$ be a representing formula. We say φ is $\mathcal{T}$-*ordinary$_G$* if $\neg\varphi(\ulcorner\varphi\urcorner) \in \mathcal{T}$. Equivalently, φ is $\mathcal{T}$-*ordinary$_G$* if it is in the set represented by its negation in $\mathcal{T}$.

Tarski's Theorem was an immediate consequence of two Lemmas, called Lemma A and Lemma B in Chapter 5, Section 8. We want to find analogs for these. We begin with Lemma A, and we suggest you go back and look at the earlier proof before reading what is here. The earlier Lemma A started by supposing the set of ordinary formulas was representable, and this quickly led to a contradiction. Well, let's try this again. We begin by introducing a few useful terms.

Definition 5.2. Let $\mathcal{T}$ be a theory.

1. $\mathcal{T}$ is *inconsistent* if there is a sentence X such that both X and $\neg X$ are in $\mathcal{T}$.

2. $\mathcal{T}$ is *incomplete* if there is a sentence X such that neither X nor $\neg X$ are in $\mathcal{T}$.

By Exercise 3.1, an inconsistent theory actually contains every sentence, and so is useless. A theory that is incomplete, on the other hand, leaves open the status of some sentences and so is of limited use. Now we turn to our analog of the proof of the earlier Lemma A. We leave until afterwards the formulation of what the proof proves.

Proof Let $\mathcal{T}$ be a Δ_0 complete theory, and suppose the set of $\mathcal{T}$-ordinary$_G$ formulas is representable in $\mathcal{T}$, say by $A(v_0)$. This means that if φ is any representing formula,

$$A(\ulcorner\varphi\urcorner) \in \mathcal{T} \iff \varphi \text{ is } \mathcal{T}\text{-ordinary}_G.$$

But, according to the definition,

$$\varphi \text{ is } \mathcal{T}\text{-ordinary}_G \iff \neg\varphi(\ulcorner\varphi\urcorner) \in \mathcal{T}$$

and so

$$A(\ulcorner\varphi\urcorner) \in \mathcal{T} \iff \neg\varphi(\ulcorner\varphi\urcorner) \in \mathcal{T}.$$

This is the case for any representing formula φ. Take φ to be A itself. Then we have

$$A(\ulcorner A\urcorner) \in \mathcal{T} \iff \neg A(\ulcorner A\urcorner) \in \mathcal{T}.$$

There are two possible ways this equivalence can hold. First, we could have that *both* $A(\ulcorner A\urcorner)$ and $\neg A(\ulcorner A\urcorner)$ are in $\mathcal{T}$. In this case $\mathcal{T}$ is *inconsistent*. Second, we could have that *neither* $A(\ulcorner A\urcorner)$ nor $\neg A(\ulcorner A\urcorner)$ are in $\mathcal{T}$. In this case $\mathcal{T}$ is *incomplete*. ∎

Let us summarize what has been proved.

Lemma A Let $\mathcal{T}$ be a Δ_0 complete theory. If the set of $\mathcal{T}$-ordinary$_G$ formulas is representable in $\mathcal{T}$, $\mathcal{T}$ is either inconsistent or incomplete.

Remark We can take T to be TS, the set of sentences of LS that are true in $\mathbb{HF}$. In this case representability in T is the same as the notion of representability used in earlier Chapters. Also, TS is clearly consistent and complete. So it follows from Lemma A in this section that TS-$ordinary_G$ is not representable, which is our earlier Lemma A in Chapter 5, Section 8.

Next we want an analog to Lemma B of Chapter 5, Section 8. We have the following, whose proof should also be compared with the earlier version.

Lemma B Let T be a Δ_0 complete formal theory. The set of T-$ordinary_G$ formulas is Σ.

Proof The set of T-$ordinary_G$ formulas consists of all formulas $\varphi(v_0)$ such that, for some closed term t naming $\varphi(v_0)$, $\neg\varphi(t) \in T$. Since T is a formal theory it is Σ, say the Σ formula $A_T(x)$ represents T. Then the set of T-$ordinary_G$ formulas is represented by: $\mathsf{Ordinary}_G(v_0) =$

$$(\exists y)(\exists x)(\exists t)\{\mathsf{RepresentingFormula}(v_0) \land \\ \mathsf{Names}(t, v_0) \land (x \text{ is } v_0(t)) \land \\ (y \text{ is } \langle\!\langle \text{`}\neg\text{'} \rangle\!\rangle * x) \land \\ A_T(y)\}$$

■

Combining these two Lemmas, we immediately have that any Δ_0 complete formal theory in which every Σ set is representable is either inconsistent or incomplete. But according to Exercise 5.1, representability of Σ sets implies consistency, so we have the following simplified version of this result.

Theorem 5.3 (Gödel's Theorem, Version 2).
Any formal Δ_0 complete theory in which every Σ set is representable is incomplete.

Corollary 5.4. *Any formal, Δ_0 complete, true theory is incomplete.*

Proof By Proposition 3.4. ■

This applies to all the theories in Section 6 of Chapter 7, but to confine ourselves to the extreme cases that were actually verified, we have the following.

Corollary 5.5. *Theory(SET_Q) and Theory(S_Σ) are incomplete.*

We were after a constructive version of Theorem 1.2 which said TS is not a formal theory. We claim Theorem 5.3 is just that. First, it does imply

Theorem 1.2, provided non-constructive methods are allowed. Technically, we have the following argument.

Corollary 5.6 (Theorem 1.2). *$\mathcal{TS}$ is not a formal theory.*

Proof $\mathcal{TS}$ is Δ_0 complete, every Σ set is representable in it, and trivially it is not incomplete. So it can not be a formal theory. ∎

Of course this argument makes essential use of $\mathcal{TS}$, which is an inherently non-finitary thing—it is an infinite set thought of as a 'completed' entity. But, the argument can be rephrased in a somewhat more concrete way, to show no formal theory could be adequate as an axiomatization of finite set theory. The reasoning is as follows. Suppose $\mathcal{T}$ is a formal, Δ_0 complete theory. If it is too weak for every Σ set to be represented in it, it is too weak to be satisfactory. And otherwise it must be incomplete, so there are questions about the hereditarily finite sets that it is unable to answer. Either way, unsatisfactory.

Notice that the incompleteness argument is entirely constructive. If every Σ set is representable in a formal theory, we can actually write out a sentence that the theory can not decide. The problem is, there is still a non-constructive aspect to all this. Suppose we are given a particular formal theory $\mathcal{T}$. Do we have a receipt for constructing an undecidable sentence in $\mathcal{T}$ or not? Well, we do provided every Σ set is representable in $\mathcal{T}$, but we have not seen any examples of theories for which this has been established to be the case constructively—in each particular case we needed to know we had a true theory, and this involves the notion of truth.

To summarize: if we have finitary evidence that a formal theory satisfies the hypotheses of Theorem 5.3, we can produce a specific example of a sentence undecidable in the theory. This argument is quite constructive. But as yet we have no example of a formal theory that has been constructively established to meet the conditions of the Theorem. We will take this point up again later on.

We conclude this section with a discussion of $\mathcal{S}_\Sigma$, which meets the conditions of Theorem 5.3, though we have no finitary proof of that fact. *Theory*($\mathcal{S}_\Sigma$) is a useful example because not only is every Σ set representable in it, but it is represented by the same Σ formulas that represents it in $\mathbb{HF}$. Speaking loosely, we know what the representing formulas are trying to say. Now, following the receipt given in the proof of Lemma A above, we can produce a sentence undecidable in *Theory*($\mathcal{S}_\Sigma$). It will be interesting to see what it asserts.

Let A be a Σ formula representing the set of $\mathcal{S}_\Sigma$-*ordinary*$_G$ formulas.

Then for each φ:

$$A(\ulcorner\varphi\urcorner) \text{ is true } \Leftrightarrow \varphi \text{ is a } Theory(\mathcal{S}_\Sigma)\text{-}ordinary_G \text{ formula.}$$

Then in particular

$$A(\ulcorner A\urcorner) \text{ is true } \Leftrightarrow A \text{ is } Theory(\mathcal{S}_\Sigma)\text{-}ordinary_G.$$

That is,
$$A(\ulcorner A\urcorner) \text{ is true } \Leftrightarrow \neg A(\ulcorner A\urcorner) \in Theory(\mathcal{S}_\Sigma).$$

Now, $A(\ulcorner A\urcorner)$ was our example of an undecidable sentence in the theory. What we have just seen is that *it asserts its own disprovability in the theory.*

In fact, since $A(\ulcorner A\urcorner)$ is undecidable, $\neg A(\ulcorner A\urcorner) \notin Theory(\mathcal{S}_\Sigma)$. Then by the last equivalence above, $A(\ulcorner A\urcorner)$ must be false (in the standard model), and so $\neg A(\ulcorner A\urcorner)$ is true. Thus $\neg A(\ulcorner A\urcorner)$ is an example of a sentence that is true in the standard model, but that is not provable from $\mathcal{S}_\Sigma$.

Notice that this is quite different than the proof of Gödel's Theorem using Tarski's Theorem 8.2, Chapter 5. In proving that result, we saw that if the set of ordinary formulas were representable, by $A(v_0)$ say, the sentence $A(\ulcorner A\urcorner)$ would have the unpleasant property of being true if and only if it was false. It followed that $A(\ulcorner A\urcorner)$ could not exist. On the other hand using the proof here, if $\mathcal{T}$ is a formal theory in which every Σ set is representable, *there will exist* a formula, also denoted $A(\ulcorner A\urcorner)$ above, that can neither be proved nor disproved in $\mathcal{T}$ (and further, $\neg A(\ulcorner A\urcorner)$ is true). The attempt to use constructive methods did give more information: we have a specific example of an undecidable sentence.

Exercises

Exercise 5.1. Show that if every Σ set is representable in a Δ_0 complete theory, the theory is consistent.

Exercise 5.2. State and prove a version of Theorem 5.3 for arithmetic.

Exercise 5.3. Let $\mathcal{T}$ be a Δ_0 complete, formal theory in the language LS. Assume instances of Σ formulas that are true in $\mathbb{HF}$ are provable in $\mathcal{T}$, and all sentences provable in $\mathcal{T}$ are true in $\mathbb{HF}$. In addition, assume the relation (x is *not* a proof of y in $\mathcal{T}$) is Σ. This is generally the case, but never mind establishing it now.

Let t be a fixed closed term. Your problem is to construct a sentence A in the language LS that "says" t is not a proof of me (in theory $\mathcal{T}$). Then show A is provable, and t is not its proof.

Here is a sketch of how to solve the problem. Call a formula $F(v_0)$ t-*ordinary* if t is not a proof that $F(v_0)$ is in the set that $F(v_0)$ represents in $\mathcal{T}$. That is, $F(v_0)$ is t-ordinary if t is not a proof of $F(f)$, where f is a closed term naming $F(v_0)$.

1. Show the collection of t-ordinary formulas is Σ. [For the remaining items, let $\Phi(v_0)$ be a Σ formula that represents the collection of t-ordinary formulas, and let $\ulcorner\Phi\urcorner$ be a closed term that names Φ.]

2. Show t is not a proof, in $\mathcal{T}$ of $\Phi(\ulcorner\Phi\urcorner)$.

3. Show $\Phi(\ulcorner\Phi\urcorner)$ is provable in $\mathcal{T}$.

Then $\Phi(\ulcorner\Phi\urcorner)$ is the sentence A that was asked for.

6 ω-Consistency

As we pointed out in the previous section, we have no examples of formal theories that meet the conditions of Gödel's Theorem 5.3, provided we insist on constructive evidence for this fact. Gödel tried to restrict the role of non-constructivity as far as possible. Whenever we have shown a formal theory to have every Σ relation representable in it, we have used models. Gödel introduced the notion of ω-consistency, which makes possible the replacement of semantical arguments by ones that only involve the notion of provability, thus gaining constructive content. Gödel worked with arithmetic—we continue to carry out our development in $\mathbb{HF}$. It makes no essential difference.

Definition 6.1. Let $\mathcal{T}$ be a theory in the language LS. $\mathcal{T}$ is ω-*inconsistent* if there is a formula $\varphi(x)$ (with only x free) such that $(\exists x)\varphi(x) \in \mathcal{T}$, but also for each closed term t, $\neg\varphi(t) \in \mathcal{T}$. The theory $\mathcal{T}$ is ω-*consistent* if it is not ω-inconsistent.

The idea is pretty simple. A theory is ω-inconsistent if it can prove something has the property φ, and at the same time it can prove each hereditarily finite set does not have the property. Notice that the notion of ω-consistency only refers to proofs—the notion of model, or of truth, does not come into it. Finally an ω-consistent theory is trivially consistent, because if it is ω-consistent something is not provable, that is, not all sentences are in $\mathcal{T}$.

To apply Theorem 5.3 we need to know every Σ set is representable in a formal theory. To show this, we need to show when certain formulas are provable, and also when they are not. (A representing formula should be provable of names for members of the set it purportedly represents, and should not be provable for names of non-members). Generally, showing something is not provable constructively is harder than showing something is. To show something is provable, we exhibit a proof, which is a finitary object. To show something is not provable, we can not generally give a counter-example unless there happens to be a finitary one. Our resources are limited and non-uniform in this direction. Gödel saw that the notion of ω-consistency could help here, essentially because the following has a simple, *constructive* proof.

Proposition 6.2. *Let $\mathcal{T}$ be a Δ_0 complete theory. If $\mathcal{T}$ is ω-consistent, every Σ set is representable in $\mathcal{T}$.*

Proof Suppose $\mathcal{T}$ is Δ_0 complete and ω-consistent. Also suppose S is a Σ set. We show S is representable in $\mathcal{T}$.

Since S is Σ, by the Normal Form Theorem 6.2 of Chapter 3 there is a Σ_1 formula that represents it in $\mathbb{HF}$, say $(\exists x)\varphi(x, v_0)$, where $\varphi(x, v_0)$ is a Δ_0 formula. We show the same formula represents S in the theory $\mathcal{T}$.

First, suppose $s \in S$. Let t be a closed term naming s. Then $(\exists x)\varphi(x, t)$ is true in $\mathbb{HF}$. But then for some closed term u, $\varphi(u, t)$ will be true in $\mathbb{HF}$. This is an instance of a Δ_0 formula, so it is provable in $\mathcal{T}$ by Δ_0 completeness. Since $\varphi(u, t) \supset (\exists x)\varphi(x, t)$ is a theorem of first-order logic, $(\exists x)\varphi(x, t)$ is provable in $\mathcal{T}$.

Conversely, suppose $(\exists x)\varphi(x, t)$ is provable in $\mathcal{T}$. Since $\mathcal{T}$ is ω-consistent, for some closed term u, $\neg\varphi(u, t)$ is not provable in $\mathcal{T}$. But $\neg\varphi(u, t)$ is also an instance of a Δ_0 formula, so it must be false in $\mathbb{HF}$, or otherwise it would have been provable in $\mathcal{T}$. Then $\varphi(u, t)$ is true in $\mathbb{HF}$, hence so is $(\exists x)\varphi(x, t)$, and so $s \in S$. ∎

While the proof above used the notion of truth extensively, if you look at it carefully you will see that it was only for Δ_0 and Σ instances. Δ_0 truth is decidable, and Σ truth has a semi-decision procedure. Consequently, the proof really is constructive in nature. Now combining this with Theorem 5.3, we have accomplished a constructive proof of the following.

Theorem 6.3 (Gödel's Theorem, Version 3). *Suppose $\mathcal{T}$ is a formal theory that is Δ_0 complete and ω-consistent. Then $\mathcal{T}$ is incomplete.*

We did our work in the context of set theory—Gödel worked in arithmetic. The techniques are the same, however. The Theorem above has exactly the same wording either way, though of course the meaning of Δ_0 shifts. Peano Arithmetic, PA, does have the property that every true instance of an arithmetic Δ_0 formula can be proved in it. In fact, this is the case for the more restricted theory Q. In both cases the argument is constructive, essentially by induction on the complexity of Δ_0 formulas. We leave the details to the dedicated. Once done, however, we have acheived an entirely constructive proof of the following, which is Gödel's Theorem as he originally stated it.

Theorem 6.4. *If Peano Arithmetic is ω-consistent, it is incomplete.*

CHAPTER 9

CHURCH'S THEOREM, ROSSER'S THEOREM

1 Introduction

There are two threads left dangling from the previous Chapter. We observed that the notion of *ordinary* split in two when we moved from representability in $\mathbb{HF}$ to representability in theories. We saw that by using one of the two versions, we arrived at Gödel's proof of his Theorem. What about the other version? This is the first of the threads we tie off in the present chapter: we will see that the other version of ordinary gives us an important result too, Church's Theorem, and so we call it *ordinary$_C$*.

The version of Gödel's Theorem we wound up with said Peano Arithmetic is incomplete provided it is ω-consistent. Now ω-consistency, while not unnatural, does seem a bit *ad hoc*. It would be nice if a simpler condition could be found to replace it and still give us a constructive argument. In fact, plain, ordinary *consistency* will do, provided Gödel's argument is replaced by a more complicated one due to Rosser. This is the second of the threads we tie off. As a matter of fact, both versions of ordinary come into play in proving Rosser's Theorem.

2 Church's Theorem

We pointed out in Chapter 8, Section 4 that the notion of ordinary which was appropriate when we were using the models $\mathbb{HF}$ and $\mathbb{N}$ split in two when extended to theories that could be incomplete or inconsistent. In Chapter 8, Section 5 we followed up the consequences of adopting one version, Definition 5.1. This led us to Gödel's proof of his First Incompleteness Theorem. Now we try the other version. As we did earlier, we carry out our argument using the language LS, rather than LA, and so we talk about Σ in preference to talking about recursively enumerable, and Δ instead of recursive. The corresponding arithmetic version is established in the same way, however.

Definition 2.1. Let T be a Δ_0 complete theory (not necessarily formal), and let $\varphi(v_0)$ be a representing formula. We call φ T-*ordinary$_C$* if $\varphi(\ulcorner\varphi\urcorner) \notin T$. Equivalently, φ is T-*ordinary$_C$* if it is not in the set it represents in T.

111

Once again we try for an analog of Lemma A from Chapter 5, Section 8. This time we come quite close.

Lemma A Let $\mathcal{T}$ be a Δ_0 complete theory. The set of $\mathcal{T}$-ordinary$_C$ formulas is not representable in $\mathcal{T}$.

Proof Suppose otherwise. Let $\mathcal{T}$ be a Δ_0 complete theory, and suppose the set of $\mathcal{T}$-ordinary$_C$ formulas is representable in $\mathcal{T}$, say by $A(v_0)$. As usual, this means that if φ is any representing formula,
$$A(\ulcorner\varphi\urcorner) \in \mathcal{T} \iff \varphi \text{ is } \mathcal{T}\text{-ordinary}_C.$$
And, according to the definition,
$$\varphi \text{ is } \mathcal{T}\text{-ordinary}_C \iff \varphi(\ulcorner\varphi\urcorner) \notin \mathcal{T}.$$
Combining these, we have
$$A(\ulcorner\varphi\urcorner) \in \mathcal{T} \iff \varphi(\ulcorner\varphi\urcorner) \notin \mathcal{T}.$$
Now take φ to be A itself. Then we have
$$A(\ulcorner A\urcorner) \in \mathcal{T} \iff A(\ulcorner A\urcorner) \notin \mathcal{T}.$$
Clearly this is impossible, and so the set of $\mathcal{T}$-ordinary$_C$ formulas is not representable in $\mathcal{T}$. ∎

Remark Once again we can see what happens if we take $\mathcal{T}$ to be $\mathcal{TS}$. Then representability in $\mathcal{T}$ is just representability in $\mathbb{HF}$, and $\mathcal{T}$-ordinary$_C$ is just plain ordinary. Thus the earlier Lemma A is a special case of this Lemma A.

Next we want an analog to Lemma B of Chapter 5, Section 8. Note that in the statement below we use the notion of Δ—it is not a misprint for Δ_0.

Lemma B Let $\mathcal{T}$ be a Δ_0 complete theory. If $\mathcal{T}$ is Δ then the set of $\mathcal{T}$-ordinary$_C$ formulas is Σ.

Proof The set of $\mathcal{T}$-ordinary$_C$ formulas consists of all representing formulas φ such that $\varphi(\ulcorner\varphi\urcorner) \notin \mathcal{T}$. If we suppose $\mathcal{T}$ is Δ, there will be a Σ formula, say $\overline{A}_{\mathcal{T}}(x)$, which represents the complement of $\mathcal{T}$. Then the set of $\mathcal{T}$-ordinary$_C$ formulas will be represented by: $\mathsf{Ordinary}_C(v_0) =$

$$(\exists x)(\exists t)\{\mathsf{RepresentingFormula}(v_0) \land \\ \mathsf{Names}(t, v_0) \land (x \text{ equals } v_0(t)) \land \\ \overline{A}_{\mathcal{T}}(x)\}$$

∎

Chapter 9. Church's Theorem, Rosser's Theorem

Combining these two Lemmas immediately gives us the following.

Theorem 2.2. *If $\mathcal{T}$ is any Δ_0 complete theory in which every Σ set is representable, the set $\mathcal{T}$ is not Δ.*

Proof If $\mathcal{T}$ is Δ, by Lemma B the set of $\mathcal{T}$-$ordinary_C$ formulas will be Σ. If also every Σ set is representable in $\mathcal{T}$, the set of $\mathcal{T}$-$ordinary_C$ formulas would be representable in $\mathcal{T}$, contradicting Lemma A. ∎

Using Propositions 1.1 and 3.4 of Chapter 8 we have the following.

Corollary 2.3. *If $\mathcal{T}$ is a Δ_0 complete, formal theory in which every Σ set is representable, then $\mathcal{T}$ is Σ but not Δ. In particular, if $\mathcal{T}$ is a Δ_0, complete, true theory, then $\mathcal{T}$ is Σ but not Δ.*

Theorem 3.6 of Chapter 6 said $\mathcal{S}_\Sigma$ itself was Σ but not Δ. This extends to the set of consequences of $\mathcal{S}_\Sigma$.

Corollary 2.4. *Theory($\mathcal{S}_\Sigma$) is Σ but not Δ.*

Church worked in an arithmetic context rather than in a set-theoretic one as we did. As we remarked above, the proof we gave works well either way, except that in the arithmetic setting Gödel numbers for formulas must be used. If this is done, the following results.

Theorem 2.5 (Church's Theorem). *Suppose $\mathcal{T}$ is an arithmetically Δ_0 complete, formal theory in which every recursively enumerable set is representable. Then $\mathcal{T}$ is recursively enumerable but not recursive.*

Whether we use the arithmetic or the set-theoretic version, Church's Theorem says that any formal theory that is sufficiently strong can have no decision procedure. The gods of mathematics must love mathematicians—they made no mechanical substitute for them.

You were asked to show, in Exercises at the end of Chapter 7, Section 6, that SET_Q is Δ_0 complete. It is also a true theory, so every Σ set is representable in SET_Q by Proposition 3.4, Chapter 8. Theorem 2.2 then says *Theory*(SET_Q) has no decision procedure. This has a remarkable consequence, because SET_Q is finitely axiomatizable. We present the following arguement informally, though it can be easily turned into a formal argument.

Suppose there was a decision procedure for first-order logic. That is, suppose we had a way of deciding which formulas are valid and which are not. Then we would have a decision procedure for *Theory*(SET_Q) as well. There are 3 axioms for SET_Q; suppose we denote them A_1, A_2, A_3. Then (using the Deduction Theorem of first-order logic) $X \in$ *Theory*(SET_Q) if and only if the sentence $(A_1 \wedge A_2 \wedge A_3) \supset X$ is valid. Clearly if we could

test for validity of first-order formulas, we could test for membership in
Theory(SET_Q). This gives us another theorem of Church.

Theorem 2.6. *There is no decision procedure for first-order logic (in the language* LS*).*

The language LS contains a relation symbol ε, a constant symbol $\varnothing$, and a function symbol $\mathbb{A}$. We saw in Section 4 of Chapter 3 that occurrences of $\varnothing$ and $\mathbb{A}$ could be eliminated in formulas in favor of ε, provided we were working in $\mathbb{HF}$ and using the semantic notion of truth. It is possible to do this with SET_Q as well. We can reformulate the axioms into the following system, SET_Q_0 (you should compare these with the axioms for SET_Q in Chapter 7, Section 6). Also recall that $x \approx y$ abbreviates a formula whose only relation symbol is ε.

1. (Extensionality) $(\forall x)(\forall y)[x \approx y \supset (\forall z)(x \, \varepsilon \, z \supset y \, \varepsilon \, z)]$

2. (Empty Set) $(\exists y)(\forall x)\neg(x \, \varepsilon \, y)$

3. (Addition) $(\forall x)(\forall y)(\exists w)(\forall z)[z \, \varepsilon \, w \equiv (z \, \varepsilon \, x \lor z \approx y)]$

Each formula X of LS can be rewritten into another formula X_0 with no constant or function symbols, just the two-place relation symbol ε, so that X is a theorem of SET_Q if and only if X_0 is a theorem of SET_Q_0. Then we can carry out the reduction of provability in SET_Q_0 to first-order validity, just as we did above for SET_Q, but now we have a simpler language. This gives us the following, which is a theorem of Church as he actually stated it.

Theorem 2.7. *There is no decision procedure for first-order logic, using a language L with a single binary relation symbol and no constant or function symbols.*

The result cannot be improved further. It can be shown that there is a decision procedure for first-order logic provided the language L has only one-place relation symbols.

Exercises

Exercise 2.1. Modify the proofs we gave for Lemma A and Lemma B, and give a proof of Theorem 2.5.

3 Rosser's Theorem

Gödel's First Incompleteness Theorem, approached via Tarski's Theorem, had a non-constructive proof. Approached via Gödel's original proof the argument is constructive, but we need the rather peculiar hypothesis of

Chapter 9. Church's Theorem, Rosser's Theorem

ω-consistency. Rosser was able to keep the argument constructive while weakening the hypothesis to that of simple consistency, but at the cost of complicating things in a rather ingenious way. We present the argument in somewhat modified terminology, using both of our notions of ordinary. First let us recall the two notions. In what follows $\mathcal{T}$ is a Δ_0 complete theory in the language LS, and φ is a representing formula.

Repeated Definitions

1. φ is $\mathcal{T}$-ordinary$_G$ if $\neg\varphi(\ulcorner\varphi\urcorner) \in \mathcal{T}$.

2. φ is $\mathcal{T}$-ordinary$_C$ if $\varphi(\ulcorner\varphi\urcorner) \notin \mathcal{T}$.

Gödel's proof turned on representing in $\mathcal{T}$ the set of $\mathcal{T}$-ordinary$_G$ formulas. The notion of ω-consistency comes up in verifying $\mathcal{T}$ meets the conditions necessary for Gödel's proof to be applicable. Church's proof, on the other hand, turned on showing the set of $\mathcal{T}$-ordinary$_C$ formulas could not be $\mathcal{T}$ represented.

It is easy to see that if $\mathcal{T}$ is consistent, any formula that is $\mathcal{T}$-ordinary$_G$ must also be $\mathcal{T}$-ordinary$_C$. Rosser's proof turns on representing in $\mathcal{T}$ a set *between* the set of $\mathcal{T}$-ordinary$_G$ formulas and the set of $\mathcal{T}$-ordinary$_C$ formulas. We have the following version of our familiar Lemma A, which combines aspects of Gödel's version and Church's version.

Lemma A Suppose $\mathcal{T}$ is a Δ_0 complete, consistent theory. If there is a set S representable in $\mathcal{T}$

$$\{\varphi \mid \varphi \text{ is } \mathcal{T}\text{-ordinary}_G\} \subseteq S \subseteq \{\varphi \mid \varphi \text{ is } \mathcal{T}\text{-ordinary}_C\}$$

then $\mathcal{T}$ is incomplete.

Proof Suppose the set S is between the sets $\{\varphi \mid \varphi \text{ is } \mathcal{T}\text{-ordinary}_G\}$ and $\{\varphi \mid \varphi \text{ is } \mathcal{T}\text{-ordinary}_C\}$ and is representable in $\mathcal{T}$, by $S(v_0)$. This means that, if φ is any representing formula,

$$\varphi \text{ is } \mathcal{T}\text{-ordinary}_G \implies S(\ulcorner\varphi\urcorner) \in \mathcal{T}$$

and

$$S(\ulcorner\varphi\urcorner) \in \mathcal{T} \implies \varphi \text{ is } \mathcal{T}\text{-ordinary}_C$$

Then according to the definitions,

$$\neg\varphi(\ulcorner\varphi\urcorner) \in \mathcal{T} \implies S(\ulcorner\varphi\urcorner) \in \mathcal{T}$$

and

$$S(\ulcorner\varphi\urcorner) \in \mathcal{T} \implies \varphi(\ulcorner\varphi\urcorner) \notin \mathcal{T}$$

This is the case for any representing formula φ. Take φ to be S itself. Then we have
$$\neg S(\ulcorner S \urcorner) \in \mathcal{T} \implies S(\ulcorner S \urcorner) \in \mathcal{T}$$
and
$$S(\ulcorner S \urcorner) \in \mathcal{T} \implies S(\ulcorner S \urcorner) \notin \mathcal{T}.$$
If we had $S(\ulcorner S \urcorner) \in \mathcal{T}$, the second of these implications would immediately give us a contradiction, so $S(\ulcorner S \urcorner) \notin \mathcal{T}$. But then it follows from the first of the implications that $\neg S(\ulcorner S \urcorner) \notin \mathcal{T}$ too. Hence $\mathcal{T}$ is incomplete. ∎

4 Rosser's Theorem Continued

Using Rosser's ideas we will be able to establish the incompleteness of theories in the language LS without using ω-consistency, just consistency. The cost, in part, is that additional assumptions about the strength of the theory must be added, but unlike ω-consistency, these are all assumptions that certain formulas *are* in the theory, and so are subject to constructive verification.

Throughout this section we use the convention that $G(x, y)$ is a formula that represents the Gödel numbering relation, x is the Gödel number of y. Also $(x \approx y)$ abbreviates the Δ_0 formula $(\forall z \,\varepsilon\, x)(z \,\varepsilon\, y) \wedge (\forall z \,\varepsilon\, y)(z \,\varepsilon\, x)$, as usual when working with LS.

Definition 4.1.   A formula $\varphi(v_0, v_1)$ *enumerates* the set $S \subseteq R_\omega$ in the Δ_0 complete theory $\mathcal{T}$ if:

1. $s \in S$ implies that for some number $n \in \omega$, $\varphi(\ulcorner s \urcorner, \ulcorner n \urcorner) \in \mathcal{T}$,

2. $s \notin S$ implies that for every number $n \in \omega$, $\neg\varphi(\ulcorner s \urcorner, \ulcorner n \urcorner) \in \mathcal{T}$.

A set is *enumerable in* $\mathcal{T}$ if some formula enumerates it.

Informally, if S is enumerable in $\mathcal{T}$ then each member of S has a numerical certificate, a number, that certifies its membership, and for non-members nothing serves as a numerical certificate of membership, and all this can be done within $\mathcal{T}$ itself.

Definition 4.2.   The formula $G(x, y)$ *strongly defines* the Gödel numbering relation in the Δ_0 complete theory $\mathcal{T}$ provided:

1. If n is the Gödel number of s then $G(\ulcorner n \urcorner, \ulcorner s \urcorner) \in \mathcal{T}$;

2. If n is not the Gödel number of s then $\neg G(\ulcorner n \urcorner, \ulcorner s \urcorner) \in \mathcal{T}$;

3. If n is the Gödel number of s then $(\forall x)[G(\ulcorner n \urcorner, x) \supset x \approx \ulcorner s \urcorner] \in \mathcal{T}$.

Chapter 9. Church's Theorem, Rosser's Theorem   117

Here is the connection between enumeration and having a strongly defined Gödel numbering relation.

Proposition 4.3. *Suppose $\mathcal{T}$ is a Δ_0 complete theory in the language* LS *and $G(x,y)$ strongly defines the Gödel numbering relation in $\mathcal{T}$. Then every Σ set is enumerable in $\mathcal{T}$.*

Proof Assume $\mathcal{T}$ is Δ_0 complete, and $G(x,y)$ strongly defines the Gödel numbering relation in $\mathcal{T}$. Let S be Σ; we show S is enumerable in $\mathcal{T}$.

By the Normal Form Theorem 6.2, Chapter 3, S is also Σ_1, so there is a Δ_0 formula $\varphi(v_0, v_1)$ such that $(\exists v_1)\varphi(v_0, v_1)$ represents S. Let $A(v_0, v_1)$ be the formula $(\exists v_2)[\varphi(v_0, v_2) \wedge G(v_1, v_2)]$. We claim this enumerates S in $\mathcal{T}$.

Suppose $s \in S$. Then $(\exists v_1)\varphi(\ulcorner s \urcorner, v_1)$ is true in $\mathbb{HF}$ since $(\exists v_1)\varphi(v_0, v_1)$ represents S. Then for some set t, $\varphi(\ulcorner s \urcorner, \ulcorner t \urcorner)$ is true. Since this is a true Δ_0 formula, $\varphi(\ulcorner s \urcorner, \ulcorner t \urcorner) \in \mathcal{T}$. Let n be the Gödel number of t. Since $G(x,y)$ strongly defines the Gödel numbering relation, $G(\ulcorner n \urcorner, \ulcorner t \urcorner) \in \mathcal{T}$. Then $[\varphi(\ulcorner s \urcorner, \ulcorner t \urcorner) \wedge G(\ulcorner n \urcorner, \ulcorner t \urcorner)] \in \mathcal{T}$, hence $(\exists v_1)[\varphi(\ulcorner s \urcorner, v_1) \wedge G(\ulcorner n \urcorner, v_1)] \in \mathcal{T}$. That is, $A(\ulcorner s \urcorner, \ulcorner n \urcorner) \in \mathcal{T}$.

Now suppose $s \notin S$, and let k be any number; we show $\neg A(\ulcorner s \urcorner, \ulcorner k \urcorner) \in \mathcal{T}$. Let t be the set that has k as its Gödel number. Since $s \notin S$, then $\varphi(\ulcorner s \urcorner, \ulcorner t \urcorner)$ is false in $\mathbb{HF}$, because otherwise we would have $(\exists v_1)\varphi(\ulcorner s \urcorner, v_1)$ true, and hence $s \in S$. Since $\neg\varphi(\ulcorner s \urcorner, \ulcorner t \urcorner)$ is a true Δ_0 formula, $\neg\varphi(\ulcorner s \urcorner, \ulcorner t \urcorner) \in \mathcal{T}$. Since $G(x,y)$ strongly defines the Gödel numbering relation in $\mathcal{T}$, $(\forall v_2)[G(\ulcorner k \urcorner, v_2) \supset v_2 \approx \ulcorner t \urcorner] \in \mathcal{T}$. It follows that $(\forall v_2)[G(\ulcorner k \urcorner, v_2) \supset \neg\varphi(\ulcorner s \urcorner, v_2)] \in \mathcal{T}$, equivalently, $\neg(\exists v_2)[G(\ulcorner k \urcorner, v_2) \wedge \varphi(\ulcorner s \urcorner, v_2)] \in \mathcal{T}$, so $\neg A(\ulcorner s \urcorner, \ulcorner k \urcorner) \in \mathcal{T}$. ∎

Next we introduce abbreviations to aid with formula readability.

$(\forall x \leq \ulcorner n \urcorner)\varphi$ abbreviates $(\forall x)[(\mathsf{Number}(x) \wedge x \leq \ulcorner n \urcorner) \supset \varphi]$
$(\exists x \leq \ulcorner n \urcorner)\varphi$ abbreviates $(\exists x)[\mathsf{Number}(x) \wedge x \leq \ulcorner n \urcorner \wedge \varphi]$

The following is central but not hard—it is left to you.

Proposition 4.4. *Assume $\mathcal{T}$ is a Δ_0 complete, formal theory. Suppose that for some number n,*

$$(\forall x \leq \ulcorner n \urcorner)[x \approx \ulcorner 0 \urcorner \vee x \approx \ulcorner 1 \urcorner \vee \ldots \vee x \approx \ulcorner n \urcorner] \in \mathcal{T}.$$

Then for every formula $\varphi(x)$

$$(\forall x \leq \ulcorner n \urcorner)\varphi(x) \equiv [\varphi(\ulcorner 0 \urcorner) \wedge \varphi(\ulcorner 1 \urcorner) \wedge \ldots \wedge \varphi(\ulcorner n \urcorner)] \in \mathcal{T}$$
$$(\exists x \leq \ulcorner n \urcorner)\varphi(x) \equiv [\varphi(\ulcorner 0 \urcorner) \vee \varphi(\ulcorner 1 \urcorner) \vee \ldots \vee \varphi(\ulcorner n \urcorner)] \in \mathcal{T}$$

Now for our version of Rosser's Theorem, slightly modified to fit our set-based approach.

Theorem 4.5 (Rosser's Theorem). *Suppose $\mathcal{T}$ is a Δ_0 complete, formal theory that meets the following conditions.*

1. *$G(x, y)$ strongly defines the Gödel numbering relation in $\mathcal{T}$.*

2. *For each number n, $(\forall x \leq \ulcorner n \urcorner)[x \approx \ulcorner 0 \urcorner \vee x \approx \ulcorner 1 \urcorner \vee \ldots \vee x \approx \ulcorner n \urcorner] \in \mathcal{T}$.*

3. *For each number n, $(\forall x)[\mathsf{Number}(x) \supset (x \leq \ulcorner n \urcorner \vee \ulcorner n \urcorner \leq x)] \in \mathcal{T}$.*

If $\mathcal{T}$ is consistent, then $\mathcal{T}$ is incomplete.

Proof Assume the various conditions on $\mathcal{T}$ given in the Proposition, including consistency. To show $\mathcal{T}$ is incomplete, according to Lemma A in the previous section it is enough to show there is a set, representable in $\mathcal{T}$, between the set of $\mathcal{T}$-$ordinary_G$ formulas and the set of $\mathcal{T}$-$ordinary_C$ formulas, and this is what we do.

Since $\mathcal{T}$ is a Δ_0 complete formal theory, the set of $\mathcal{T}$-$ordinary_G$ formulas is Σ—this is Lemma B in Chapter 8, Section 5. A similar argument shows the set of formulas that are *not* $\mathcal{T}$-$ordinary_C$ is also Σ. Then by Proposition 4.3, both these sets are enumerable in $\mathcal{T}$. Let us say $A(x, y)$ enumerates in $\mathcal{T}$ the set of $\mathcal{T}$-$ordinary_G$ formulas, and $B(x, y)$ enumerates in $\mathcal{T}$ the set of formulas that are not $\mathcal{T}$-$ordinary_C$.

Now, let $S(v_0)$ be the following formula.

$$(\forall y)\{\mathsf{Number}(y) \supset [B(v_0, y) \supset (\exists z)[\mathsf{Number}(z) \wedge z \leq y \wedge A(v_0, z)]]\}.$$

We claim that in $\mathcal{T}$ the set that $S(v_0)$ represents is intermediate between the set of $\mathcal{T}$-$ordinary_G$ formulas and the set of $\mathcal{T}$-$ordinary_C$ formulas, which will complete the proof.

Part I Suppose F is $\mathcal{T}$-$ordinary_G$; we show it is in the set that $S(v_0)$ represents in $\mathcal{T}$.

F is $\mathcal{T}$-$ordinary_G$, and $A(x, y)$ enumerates the collection of $\mathcal{T}$-$ordinary_G$ formulas in $\mathcal{T}$. Then for some number n, $A(\ulcorner F \urcorner, \ulcorner n \urcorner) \in \mathcal{T}$. Also by definition of $\mathcal{T}$-$ordinary_G$, $\neg F(\ulcorner F \urcorner) \in \mathcal{T}$, and $\mathcal{T}$ is consistent, so $F(\ulcorner F \urcorner) \notin \mathcal{T}$, which tells us F is $\mathcal{T}$-$ordinary_C$ as well. Since $B(x, y)$ enumerates the formulas that are not $\mathcal{T}$-$ordinary_C$, for every number k, $\neg B(\ulcorner F \urcorner, \ulcorner k \urcorner) \in \mathcal{T}$. Then $\neg B(\ulcorner F \urcorner, \ulcorner 0 \urcorner) \wedge \neg B(\ulcorner F \urcorner, \ulcorner 1 \urcorner) \wedge \ldots \wedge \neg B(\ulcorner F \urcorner, \ulcorner n \urcorner) \in \mathcal{T}$, so by Proposition 4.4, $(\forall y \leq \ulcorner n \urcorner) \neg B(\ulcorner F \urcorner, y) \in \mathcal{T}$. Unabbreviating,

$$(\forall y)[(\mathsf{Number}(y) \wedge y \leq \ulcorner n \urcorner) \supset \neg B(\ulcorner F \urcorner, y)]$$

and then by classical logic,

$$(\forall y)\{\mathsf{Number}(y) \supset [B(\ulcorner F \urcorner, y) \supset \neg(y \leq \ulcorner n \urcorner)]\}$$

so using assumption 3,

$$(\forall y)\{\mathsf{Number}(y) \supset [B(\ulcorner F \urcorner, y) \supset (\ulcorner n \urcorner \leq y)]\}$$

and since $A(\ulcorner F \urcorner, \ulcorner n \urcorner) \in \mathcal{T}$,

$$(\forall y)\{\mathsf{Number}(y) \supset [B(\ulcorner F \urcorner, y) \supset (\ulcorner n \urcorner \leq y \wedge A(\ulcorner F \urcorner, \ulcorner n \urcorner))]\}$$

hence (using the fact that the true Δ_0 formula instance $\mathsf{Number}(\ulcorner n \urcorner)$ is in $\mathcal{T}$),

$$(\forall y)\{\mathsf{Number}(y) \supset [B(\ulcorner F \urcorner, y) \supset (\exists z)[\mathsf{Number}(z) \wedge z \leq y \wedge A(\ulcorner F \urcorner, z)]]\}$$

or $S(\ulcorner F \urcorner)$. This complete the first half of the argument.

Part II Suppose F in the set that $S(v_0)$ represents in $\mathcal{T}$; we show F is $\mathcal{T}$-ordinary$_C$. The argument is in the contrapositive direction.

Suppose F is not $\mathcal{T}$-ordinary$_C$. Then since $B(x, y)$ enumerates the non $\mathcal{T}$-ordinary$_C$ formulas, for some number n, $B(\ulcorner F \urcorner, \ulcorner n \urcorner) \in \mathcal{T}$. Since $\mathcal{T}$ is consistent every $\mathcal{T}$-ordinary$_G$ formula is $\mathcal{T}$-ordinary$_C$ so it follows that F is not $\mathcal{T}$-ordinary$_G$, and so for every number k, $\neg A(\ulcorner F \urcorner, \ulcorner k \urcorner) \in \mathcal{T}$. In particular, $\neg A(\ulcorner F \urcorner, \ulcorner 0 \urcorner) \wedge \neg A(\ulcorner F \urcorner, \ulcorner 1 \urcorner) \wedge \ldots \wedge \neg A(\ulcorner F \urcorner, \ulcorner n \urcorner) \in \mathcal{T}$, so $(\forall z \leq \ulcorner n \urcorner)\neg A(\ulcorner F \urcorner, z) \in \mathcal{T}$, by Proposition 4.4. We thus have

$$B(\ulcorner F \urcorner, \ulcorner n \urcorner) \wedge (\forall z \leq \ulcorner n \urcorner)\neg A(\ulcorner F \urcorner, z) \in \mathcal{T}$$

But then we have

$$\neg[B(\ulcorner F \urcorner, \ulcorner n \urcorner) \supset \neg(\forall z \leq \ulcorner n \urcorner)\neg A(\ulcorner F \urcorner, z)] \in \mathcal{T}$$

from which follows

$$\neg[B(\ulcorner F \urcorner, \ulcorner n \urcorner) \supset (\exists z)[\mathsf{Number}(z) \wedge z \leq \ulcorner n \urcorner \wedge A(\ulcorner F \urcorner, z)]] \in \mathcal{T}.$$

Then, making use of Δ_0 completeness,

$$\neg(\forall y)\{\mathsf{Number}(y) \supset [B(\ulcorner F \urcorner, y) \supset$$
$$(\exists z)[\mathsf{Number}(z) \wedge z \leq y \wedge A(\ulcorner F \urcorner, z)]]\} \in \mathcal{T}$$

that is, $\neg S(\ulcorner F \urcorner) \in \mathcal{T}$. Since $\mathcal{T}$ is consistent, $S(\ulcorner F \urcorner) \notin \mathcal{T}$. ∎

The proof is finished and, we must admit, it is rather technical. It would be good to go through it again from a slightly different point of view. We do so next—think of it as an alternative argument for Theorem 4.5.

In the proof above we constructed a particular formula,

$$S(v_0) = (\forall y)\{\mathsf{Number}(y) \supset [B(v_0,y) \supset (\exists z)[\mathsf{Number}(z) \land z \leq y \land A(v_0,z)]]\}.$$

We have seen enough of diagonal arguments by now to realize we should look carefully at the formula $S(\ulcorner S \urcorner)$.

$$S(\ulcorner S \urcorner) = (\forall y)\{\mathsf{Number}(y) \supset [B(\ulcorner S \urcorner, y) \supset \\ (\exists z)[\mathsf{Number}(z) \land z \leq y \land A(\ulcorner S \urcorner, z)]]\}$$

What does $S(\ulcorner S \urcorner)$ 'say'? We know $A(v_0, v_1)$ enumerates the $\mathcal{T}\text{-}ordinary_G$ formulas and $B(v_0, v_1)$ enumerates the non $\mathcal{T}\text{-}ordinary_C$ formulas. Then, in $\mathcal{T}$, $A(\ulcorner S \urcorner, \ulcorner n \urcorner)$ 'says' that n is evidence that S is $\mathcal{T}\text{-}ordinary_G$, and $B(\ulcorner S \urcorner, \ulcorner n \urcorner)$ 'says' that n is evidence that S is not $\mathcal{T}\text{-}ordinary_C$. We also know from Chapter 1, Section 5 that more complex sets, that is sets with higher rank, have larger Gödel numbers. Then, loosely, $S(\ulcorner S \urcorner)$ says: if there is evidence that I am not $\mathcal{T}\text{-}ordinary_C$, there is simpler evidence (that is, a smaller number) that I am $\mathcal{T}\text{-}ordinary_G$.

Suppose we had $S(\ulcorner S \urcorner) \in \mathcal{T}$. Then S would not be $\mathcal{T}\text{-}ordinary_C$, so for some number n, $B(\ulcorner S \urcorner, \ulcorner n \urcorner) \in \mathcal{T}$. Using universal instantiation on $S(\ulcorner S \urcorner)$, which is in $\mathcal{T}$, we have the following.

$$\{\mathsf{Number}(\ulcorner n \urcorner) \supset [B(\ulcorner S \urcorner, \ulcorner n \urcorner) \\ \supset (\exists z)[\mathsf{Number}(z) \land z \leq \ulcorner n \urcorner \land A(\ulcorner S \urcorner, z)]]\} \in \mathcal{T}$$

By Δ_0 completeness $\mathsf{Number}(\ulcorner n \urcorner) \in \mathcal{T}$ so by *modus ponens* we have the following.

$$(\exists z)[\mathsf{Number}(z) \land z \leq \ulcorner n \urcorner \land A(\ulcorner S \urcorner, z)] \in \mathcal{T}$$

But then, by Proposition 4.4

$$A(\ulcorner S \urcorner, \ulcorner 0 \urcorner) \lor A(\ulcorner S \urcorner, \ulcorner 1 \urcorner) \lor \ldots \lor A(\ulcorner S \urcorner, \ulcorner n \urcorner) \in \mathcal{T}.$$

We are assuming that $\mathcal{T}$ is consistent, so since $S(\ulcorner S \urcorner) \in \mathcal{T}$ then $\neg S(\ulcorner S \urcorner) \notin \mathcal{T}$, and consequently $S(v_0)$ is not $\mathcal{T}\text{-}ordinary_G$. But then for every integer k, $\neg A(\ulcorner S \urcorner, \ulcorner k \urcorner) \in \mathcal{T}$. Consequently

$$\neg A(\ulcorner S \urcorner, \ulcorner 0 \urcorner) \land \neg A(\ulcorner S \urcorner, \ulcorner 1 \urcorner) \land \ldots \land \neg A(\ulcorner S \urcorner, \ulcorner n \urcorner) \in \mathcal{T}$$

and we have that $\mathcal{T}$ is inconsistent. We conclude that $S(\ulcorner S \urcorner) \notin \mathcal{T}$.

Finally, suppose we had $\neg S(\ulcorner S \urcorner) \in \mathcal{T}$. It follows that $S(v_0)$ would be $\mathcal{T}$-ordinary$_G$, so for some number n, $A(\ulcorner S \urcorner, \ulcorner n \urcorner) \in \mathcal{T}$. Then we have the following.

$$(\forall y)\{\mathsf{Number}(y) \supset [\ulcorner n \urcorner \leq y \supset (\exists z)[\mathsf{Number}(z) \wedge z \leq y \wedge A(\ulcorner S \urcorner, z)]]\} \in \mathcal{T}$$

We have already showed that $S(\ulcorner S \urcorner) \notin \mathcal{T}$, so for every number k, $\neg B(\ulcorner S \urcorner, \ulcorner k \urcorner) \in \mathcal{T}$. Consequently

$$\neg B(\ulcorner S \urcorner, \ulcorner 0 \urcorner) \wedge \neg B(\ulcorner S \urcorner, \ulcorner 1 \urcorner) \wedge \ldots \wedge \neg B(\ulcorner S \urcorner, \ulcorner n \urcorner) \in \mathcal{T}$$

and so

$$(\forall y \leq \ulcorner n \urcorner) \neg B(\ulcorner S \urcorner, y) \in \mathcal{T}.$$

Unabbreviating this, and doing a little classical manipulation, we have

$$(\forall y)\{\mathsf{Number}(y) \supset [y \leq \ulcorner n \urcorner \supset \neg B(\ulcorner S \urcorner, y)]\} \in \mathcal{T}.$$

Since we are assuming $(\forall x)[\mathsf{Number}(x) \supset (x \leq \ulcorner n \urcorner \vee \ulcorner n \urcorner \leq x)] \in \mathcal{T}$, we conclude

$$(\forall y)\{\mathsf{Number}(y) \supset [\neg B(\ulcorner S \urcorner, y) \vee (\exists z)[\mathsf{Number}(z) \wedge z \leq y \wedge A(\ulcorner S \urcorner, z)]]\} \in \mathcal{T}.$$

But this is equivalent to $S(\ulcorner S \urcorner) \in \mathcal{T}$, which we know is not the case. We conclude that $\neg S(\ulcorner S \urcorner) \notin \mathcal{T}$ either, and so $\mathcal{T}$ is incomplete.

Exercises

Exercise 4.1. Show that condition 2 of Definition 4.2 follows from conditions 1 and 3.

Exercise 4.2. Give a proof of Proposition 4.4.

CHAPTER 10

GÖDEL'S SECOND THEOREM

1 Introduction

We have proved the first of Gödel's incompleteness theorems several times now. We gave a semantic argument making use of Tarski's Theorem in Section 1 of Chapter 8. We gave a proof closer in spirit to the original one of Gödel in Section 5 of that Chapter. We gave Rosser's version starting in Chapter 9, Section 3. Other arguments are available, for instance one can derive Gödel's result from the unsolvability of the Halting Problem, Turing's Theorem. But for Gödel's Second Incompleteness Theorem there is essentially only one argument that is common in the literature, Gödel's original one. At its heart is an elegant idea, but getting to it involves considerable messy detail. The messy detail all goes toward proving essential results that one is probably willing to believe anyway, so it is common to omit the details, state these results, and proceed from there. In fact this is what Gödel himself did—the details were filled in later by Hilbert and Bernays—and this is what we will do here as well.

2 The Gödel Fixed Point Theorem

In Chapter 5 we proved a fixed point result, Theorem 9.2, but we did so semantically. Some of the 'messy details' referred to in the Introduction come from the fact that now the work must be carried out in a formal system, with provability replacing the notion of truth. What follows is fairly close to Gödel's original argument, except that we work in the language *LS* of set theory, whereas Gödel worked in arithmetic. And rather than use a specific formal theory, we abstract as much as we can.

Definition 2.1. Let T be a Δ_0 complete theory in the language *LS*. We say T *has the fixpoint property* provided, for every formula $\varphi(v_0)$ with one free variable there is a sentence X such that $\varphi(\ulcorner X \urcorner) \equiv X$ is in T.

We need some general conditions on a theory which will ensure it has the fixpoint property, and for this some special terminology is useful. If $F(x_1, \ldots, x_n, y)$ is a formula in the language *LS*, of course F defines y as a function of $x_1, \ldots, x_n$ in $\mathbb{HF}$ if the relation that the formula represents is single-valued. We need a version of this that can be applied within formal theories, but as weak as possible so that it is not too hard for us to verify that a theory meets the condition. The following is an analog of a notion called *completely definable* in arithmetic approaches to this subject.

Definition 2.2. Let $F(x_1,\ldots,x_n,y)$ be a formula in the language LS. We say F defines y in a *function-like* way from x_1, ..., x_n in a theory $\mathcal{T}$ provided, for closed terms t_1, ..., t_n and u, if $F(t_1,\ldots,t_n,u)$ is in $\mathcal{T}$ so is the following.

$$(\forall y)\{F(t_1,\ldots,t_n,y) \supset [y \approx u]\}$$

Theorem 2.3 (Gödel's Fixed Point Theorem). *Let $\mathcal{T}$ be a Δ_0 complete theory in the language LS meeting the conditions:*

1. *$\mathcal{T}$ contains every true closed instance of a Σ formula;*

2. *The Σ formula $\mathsf{Names}(y,x)$ defines y in a function-like way from x in $\mathcal{T}$, where $\mathsf{Names}(y,x)$ represents the relation 'y is a closed term that names the set x';*

3. *The Σ formula $(z \text{ is } x(u))$ defines z in a function-like way from x and u in $\mathcal{T}$, where $(z \text{ is } x(u))$ represents the relation 'x is a representing formula and z is the result of substituting u for v_0 in x'.*

Given all this, $\mathcal{T}$ has the fixpoint property.

A remark: Instead of Σ formulas we could restrict things to Σ_1 formulas, in which case condition 1 would follow from Δ_0 completeness using Proposition 5.7, Chapter 7. Many formal theories are strong enough to show, within the theory, that Σ formulas can be reduced to Σ_1 formulas, but we do not pursue that here.

Proof The proof of Theorem 9.2 in Chapter 5 was constructive, and began as follows. Let $\varphi(x)$ be a formula with one free variable. Define $A(v_0)$ as follows.

$$\begin{aligned}A(v_0) = (\exists x)(\exists t)\{&\mathsf{RepresentingFormula}(v_0) \wedge \\ &\mathsf{Names}(t,v_0) \wedge (x \text{ is } v_0(t)) \wedge \\ &\varphi(x)\}\end{aligned}$$

Then, if we set $X = A(\ulcorner A \urcorner)$ the formula $\varphi(\ulcorner X \urcorner) \equiv X$ is true. We will show $\varphi(\ulcorner X \urcorner) \equiv X$ is also in the theory $\mathcal{T}$, which establishes the present fixed point theorem. The argument is in two parts, one for each direction of the implication. First we deal with some matters that are needed for both parts.

Chapter 10. Gödel's Second Theorem

Let a be a specific closed term of LS that names the formula $A(v_0)$ in the standard model $\mathbb{HF}$, and let X be $A(a)$. Also let $\mathbf{X}$ be a specific closed term of LS that names X. We will show $\varphi(\mathbf{X}) \equiv X$ is in $\mathcal{T}$.

Since $A(v_0)$ is a representing formula, and the closed term a names it, RepresentingFormula(a) is true in $\mathbb{HF}$, and since it is a true instance of a Σ formula, it is in $\mathcal{T}$. Also a is a set; let $\mathbf{a}$ be a closed term that names a. Since a names $A(v_0)$, and $\mathbf{a}$ and a are terms naming a and $A(v_0)$ respectively, Names$(\mathbf{a}, a)$ is true in $\mathbb{HF}$ and, also being a true instance of a Σ formula, is in $\mathcal{T}$. X is the formula $A(a)$. $\mathbf{X}$ is a closed term that names X. Then, since a names $A(v_0)$, $\mathbf{a}$ names a, and $\mathbf{X}$ names X, that is $A(a)$, the formula ($\mathbf{X}$ is $a(\mathbf{a})$) is true in $\mathbb{HF}$, and hence also is in $\mathcal{T}$.

Proof that $\varphi(\mathbf{X}) \supset X$ is in $\mathcal{T}$. Using items above we have, in $\mathcal{T}$.

$$\varphi(\mathbf{X}) \supset \{\text{RepresentingFormula}(a) \land$$
$$\text{Names}(\mathbf{a}, a) \land (\mathbf{X} \text{ is } a(\mathbf{a})) \land$$
$$\varphi(\mathbf{X})\}$$
$$\supset (\exists x)(\exists t)\{\text{RepresentingFormula}(a) \land$$
$$\text{Names}(t, a) \land (x \text{ is } a(t)) \land$$
$$\varphi(x)\}$$

But the consequent is $A(a)$, that is, X.

Proof that $X \supset \varphi(\mathbf{X})$ is in $\mathcal{T}$. Since Names$(\mathbf{a}, a)$ is in $\mathcal{T}$, by hypothesis 2 the following is in $\mathcal{T}$:

$$(\forall t)\{\text{Names}(t, a) \supset (t \approx \mathbf{a})\} \tag{10.1}$$

Since ($\mathbf{X}$ is $a(\mathbf{a})$) is in $\mathcal{T}$, by hypothesis 3 the following is in $\mathcal{T}$:

$$(\forall x)\{(x \text{ is } a(\mathbf{a})) \supset (x \approx \mathbf{X})\} \tag{10.2}$$

Now in $\mathcal{T}$ we have the following.

$$X = A(a) = (\exists x)(\exists t)\{\text{RepresentingFormula}(a) \land$$
$$\text{Names}(t, a) \land (x \text{ is } a(t)) \land \varphi(x)\} \tag{10.3}$$
$$\supset (\exists t)\{\text{Names}(t, a) \land (\exists x)[(x \text{ is } a(t)) \land \varphi(x)]\} \tag{10.4}$$
$$\supset (\exists x)(x \text{ is } a(\mathbf{a}) \land \varphi(x))] \tag{10.5}$$
$$\supset \varphi(\mathbf{X}) \tag{10.6}$$

(10.3) is by definition and (10.4) is a rearrangement and weakening. (10.5) uses (10.1), (10.6) is similar using (10.2). ∎

Incidentally, we never required that $\mathcal{T}$ be a *formal* theory, so it could be the set of sentences of LS that are true in $\mathbb{HF}$. Using this choice, Theorem 2.3 yields Theorem 9.2, Chapter 5, because the hypotheses of Theorem 2.3 are trivially true for $\mathbb{HF}$.

3 The Löb Provability Conditions

The key to proving Gödel's Second Incompleteness Theorem for a theory $\mathcal{T}$ is to show that the proof of Gödel's First Incompleteness Theorem can be carried out within $\mathcal{T}$ itself. For the arithmetic theory that Gödel originally worked with (not exactly Peano arithmetic, but a close relative), Gödel simply asserted that this could be done. In fact doing so is surprisingly complicated, requiring much careful attention to details. In order to simplify things, Hilbert and Bernays formulated some conditions on provability that would be necessary to show, and from which Gödel's result followed rather easily. Much later, Löb provided a simpler and more natural set of conditions, and these are what we will use here. They are commonly known as the *Löb provability conditions*. We give them for set theory, rather than arithmetic, though the form is exactly the same. There are three conditions. In what follows we argue that, under quite general circumstances, the conditions are *true* in $\mathbb{HF}$. This is simple. What is difficult is showing the conditions are actually provable in particular formal theories. This is something we do not undertake to do, though detailed arguments can be found elsewhere.

In what follows we assume $\mathcal{T}$ is a formal theory, and $\mathsf{Bew}_\mathcal{T}(v_0)$ is a Σ formula that represents membership in $\mathcal{T}$—provability from the axioms for $\mathcal{T}$. (This is essentially Gödel's original notation. Bew stands for 'Beweis', which is German for 'proof'.) We also assume that $\mathcal{T}$ is Δ_0 complete and all true instances of Σ formulas are in $\mathcal{T}$.

Let X be a formula in the language LS. Suppose X is in $\mathcal{T}$, and hence $\mathsf{Bew}_\mathcal{T}(\ulcorner X \urcorner)$ is true in $\mathbb{HF}$. Since this is a true instance of a Σ formula, $\mathsf{Bew}_\mathcal{T}(\ulcorner X \urcorner)$ must, itself, be in $\mathcal{T}$. This gives us the first of the Provability Conditions.

$$\text{If } X \in \mathcal{T} \text{ then } \mathsf{Bew}_\mathcal{T}(\ulcorner X \urcorner) \in \mathcal{T}$$

Any theory is closed under *modus ponens*, from $X \supset Y$ and X to conclude Y. Consequently the following is true in $\mathbb{HF}$, for any formulas X and Y: $\mathsf{Bew}_\mathcal{T}(\ulcorner X \supset Y \urcorner) \supset (\mathsf{Bew}_\mathcal{T}(\ulcorner X \urcorner) \supset \mathsf{Bew}_\mathcal{T}(\ulcorner Y \urcorner))$. The next of the Provability Conditions is that $\mathcal{T}$ should, itself, be able to prove this, that is, this formula is in $\mathcal{T}$. This is not a Σ formula, so showing particular formal theories have this property takes work, but it is more-or-less straightforward.

$$\mathsf{Bew}_\mathcal{T}(\ulcorner X \supset Y \urcorner) \supset (\mathsf{Bew}_\mathcal{T}(\ulcorner X \urcorner) \supset \mathsf{Bew}_\mathcal{T}(\ulcorner Y \urcorner)) \in \mathcal{T}$$

Suppose $\mathsf{Bew}_\mathcal{T}(\ulcorner X \urcorner)$ is true in $\mathbb{HF}$. Since this is a Σ formula, it must be in $\mathcal{T}$. But for any Z, if Z is in $\mathcal{T}$, $\mathsf{Bew}_\mathcal{T}(\ulcorner Z \urcorner)$ is true, since $\mathsf{Bew}_\mathcal{T}(v_0)$ represents $\mathcal{T}$. Then $\mathsf{Bew}_\mathcal{T}(\ulcorner \mathsf{Bew}_\mathcal{T}(\ulcorner X \urcorner)\urcorner)$ must be true in $\mathbb{HF}$. We have argued that $\mathsf{Bew}_\mathcal{T}(\ulcorner X \urcorner) \supset \mathsf{Bew}_\mathcal{T}(\ulcorner \mathsf{Bew}_\mathcal{T}(\ulcorner X \urcorner)\urcorner)$ is true in $\mathbb{HF}$. The final Provability Condition is that this formula should be in $\mathcal{T}$.

$$\mathsf{Bew}_\mathcal{T}(\ulcorner X \urcorner) \supset \mathsf{Bew}_\mathcal{T}(\ulcorner \mathsf{Bew}_\mathcal{T}(\ulcorner X \urcorner)\urcorner) \in \mathcal{T}$$

For particular theories, this condition is the hardest to establish.

Definition 3.1. We say a formal theory $\mathcal{T}$ *satisfies the Löb provability conditions* if the following holds.

1. If $X \in \mathcal{T}$ then $\mathsf{Bew}_\mathcal{T}(\ulcorner X \urcorner) \in \mathcal{T}$
2. $\mathsf{Bew}_\mathcal{T}(\ulcorner X \supset Y \urcorner) \supset (\mathsf{Bew}_\mathcal{T}(\ulcorner X \urcorner) \supset \mathsf{Bew}_\mathcal{T}(\ulcorner Y \urcorner)) \in \mathcal{T}$
3. $\mathsf{Bew}_\mathcal{T}(\ulcorner X \urcorner) \supset \mathsf{Bew}_\mathcal{T}(\ulcorner \mathsf{Bew}_\mathcal{T}(\ulcorner X \urcorner)\urcorner) \in \mathcal{T}$

We do not prove it, but *FIN_SET* from Chapter 7, Section 6.2 satisfies the Löb provability conditions.

4 Abbreviated Notation

We will now rewrite the provability conditions in a rather suggestive way. Instead of $\mathsf{Bew}_\mathcal{T}(\ulcorner X \urcorner)$, let us write just $\Box X$. For the time being this is just a convenient abbreviation, but it will become something more shortly. In this new notation the conditions can be stated as follows, with the first condition reformulated so that it looks like an inference rule. The other two should be understood as assertions that the formulas displayed are in $\mathcal{T}$.

Necessitation Rule $\dfrac{X}{\Box X}$

K Axiom Scheme $\Box(X \supset Y) \supset (\Box X \supset \Box Y)$

4 Axiom Scheme $\Box X \supset \Box\Box X$

Those of you who are familiar with such things will recognize the items above as the modal axiom schemes and rule of the standard modal logic K4. Using a modal formulation of provability traces back to Gödel, who introduced the now-standard Necessitation Rule. For our purposes we don't need to know anything about the original motivations for studying modal logic, or its various intended applications. We can simply apply the axiom schemes and rule above, of course together with classical propositional tautologies and *modus ponens*, thus reasoning in the logic K4, and the results will tell us something about the behavior of $\text{Bew}_\mathcal{T}$ for those formal theories $\mathcal{T}$ that satisfy the Löb provability conditions. Here are some K4 results that will be applied in the next section.

First there is a derived rule of K4, usually called the *regularity rule*, which says we can infer $\Box X \supset \Box Y$ from $X \supset Y$. This inference abbreviates the following steps.

1. $X \supset Y$
2. $\Box(X \supset Y)$ (from 1 by the necessitation rule)
3. $\Box(X \supset Y) \supset (\Box X \supset \Box Y)$ (**K** axiom scheme)
4. $\Box X \supset \Box Y$ (from 2 and 3 by *modus ponens*)

Next, $\Box(X \wedge Y) \supset (\Box X \wedge \Box Y)$ is provable in K4. Here is the argument.

1. $(X \wedge Y) \supset X$ (a tautology)
2. $\Box(X \wedge Y) \supset \Box X$ (from 1 by the regularity rule)
3. $\Box(X \wedge Y) \supset \Box Y$ (similarly)
4. $\Box(X \wedge Y) \supset (\Box X \wedge \Box Y)$ (from 2 and 3 by classical propositional logic)

The converse, $(\Box X \wedge \Box Y) \supset \Box(X \wedge Y)$, is also provable in K4.

1. $X \supset (Y \supset (X \wedge Y))$ (a tautology)
2. $\Box X \supset \Box(Y \supset (X \wedge Y))$ (from 1 by the regularity rule)
3. $\Box(Y \supset (X \wedge Y)) \supset (\Box Y \supset \Box(X \wedge Y))$ (**K** axiom scheme)
4. $\Box X \supset (\Box Y \supset \Box(X \wedge Y))$ (from 2 and 3 by classical logic)
5. $(\Box X \wedge \Box Y) \supset \Box(X \wedge Y)$ (from 4 by classical logic)

Thus we have $\Box(X \wedge Y) \equiv (\Box X \wedge \Box Y)$ is provable in K4. And remember, this tells us that if $\mathcal{T}$ is any formal theory that satisfies the Löb provability conditions, then $\text{Bew}_\mathcal{T}(\ulcorner X \wedge Y \urcorner) \equiv (\text{Bew}_\mathcal{T}(\ulcorner X \urcorner) \wedge \text{Bew}_\mathcal{T}(\ulcorner Y \urcorner))$ is in $\mathcal{T}$.

Exercises

Exercise 4.1. Show that if $\mathcal{T}$ meets the Löb provability conditions, then if X and Y are both classically contradictory formulas, then $\text{Bew}_\mathcal{T}(\ulcorner X \urcorner) \equiv \text{Bew}_\mathcal{T}(\ulcorner Y \urcorner) \in \mathcal{T}$.

5 Gödel's Second Incompleteness Theorem

For starters we give half the argument for the First Incompleteness Theorem along Gödel's original lines. We will be somewhat informal about it, since the argument is needed for motivation only. Assume $\mathcal{T}$ is a Δ_0 complete formal theory, containing true instances of Σ formulas, having the fixpoint property, and satisfying the Löb provability conditions.

By the Fixed Point Theorem, there is an X such that $X \equiv \neg\mathsf{Bew}_\mathcal{T}(\ulcorner X \urcorner)$ is in $\mathcal{T}$, or in the abbreviated modal notation, such that $X \equiv \neg\Box X$ is in $\mathcal{T}$. We claim X is not in the theory $\mathcal{T}$, provided $\mathcal{T}$ is consistent.

1. Assume $X \in \mathcal{T}$.

2. Then $\Box X$ is true.

3. Then $\Box X \in \mathcal{T}$, since true Σ formulas are in $\mathcal{T}$.

4. But $X \equiv \neg\Box X \in \mathcal{T}$.

5. So $\neg\Box X \in \mathcal{T}$, by 1 and 4.

6. Then $\mathcal{T}$ is inconsistent, by 3 and 5.

Reversing things, if $\mathcal{T}$ is consistent, X is not in $\mathcal{T}$. Incidentally, note that X 'says' it is not in $\mathcal{T}$, since it is equivalent to $\neg\mathsf{Bew}_\mathcal{T}(\ulcorner X \urcorner)$, hence it is true, provided $\mathcal{T}$ is consistent.

Now the idea is to carry out the argument we just did, but *within* $\mathcal{T}$ itself. Then, 'is the case' should become 'is provable in $\mathcal{T}$ to be the case.' This idea will become clearer as we proceed.

The argument above concluded, in step 6, with $\mathcal{T}$ being inconsistent because it contained a contradiction. We need some way of saying $\mathcal{T}$ is inconsistent within $\mathcal{T}$ itself. To this end, let $\bot$ be your favorite contradiction, $Z \wedge \neg Z$ for some Z, say. Then, $\mathsf{Bew}_\mathcal{T}(\ulcorner \bot \urcorner)$ expresses the inconsistency of $\mathcal{T}$. (The choice of a specific contradiction won't matter, by Exercise 4.1.) And then, $\neg\mathsf{Bew}_\mathcal{T}(\ulcorner \bot \urcorner)$ expresses the consistency of $\mathcal{T}$.

Theorem 5.1 (Gödel's Second Incompleteness Theorem). *Let $\mathcal{T}$ be a Δ_0 complete formal theory, containing true instances of Σ formulas, having the fixpoint property, and satisfying the Löb provability conditions. If $\mathcal{T}$ is consistent then $\neg\mathsf{Bew}_\mathcal{T}(\ulcorner \bot \urcorner)$ does not belong to $\mathcal{T}$.*

Briefly, a sufficiently expressive theory that is consistent cannot prove its own consistency.

Proof We make use of abbreviated notation, writing $\Box Z$ for $\mathsf{Bew}_{\mathcal{T}}(\ulcorner Z \urcorner)$. By Theorem 2.3 there is a formula X that is a fixed point for $\neg\mathsf{Bew}_{\mathcal{T}}(v_0)$, that is, $X \equiv \neg\mathsf{Bew}_{\mathcal{T}}(\ulcorner X \urcorner)$ is in $\mathcal{T}$. Using modal abbreviations, $X \equiv \neg\Box X$ is in $\mathcal{T}$. We just showed it follows that $X \notin \mathcal{T}$. Now, here is the argument for the Second Incompleteness Theorem—you should compare it with the First Incompleteness Theorem argument given above.

1. $(X \equiv \neg\Box X) \in \mathcal{T}$ (since X is a fixed point)
2. $(X \supset \neg\Box X) \in \mathcal{T}$ (from 1)
3. $(\Box X \supset \Box\neg\Box X) \in \mathcal{T}$ (from 2 using the regularity rule)
4. $(\Box X \supset \Box\Box X) \in \mathcal{T}$ (4 axiom scheme)
5. $(\Box X \supset (\Box\neg\Box X \wedge \Box\Box X)) \in \mathcal{T}$ (from 3 and 4)
6. $(\Box X \supset \Box(\neg\Box X \wedge \Box X)) \in \mathcal{T}$ (from 5 and results in the previous section)
7. $(\Box X \supset \Box\bot) \in \mathcal{T}$ (equivalent of 6)
8. $(\neg\Box\bot \supset \neg\Box X) \in \mathcal{T}$ (contrapositive of 7)
9. $(\neg\Box\bot \supset X) \in \mathcal{T}$ (from 1 and 8)

Now, if $\neg\Box\bot$ were in $\mathcal{T}$ we would also have $X \in \mathcal{T}$. But by the Gödel *First* Incompleteness Theorem, $X \notin \mathcal{T}$. Hence $\neg\Box\bot \notin \mathcal{T}$. ∎

Exercises

Exercise 5.1. Let $\mathcal{T}$ be a theory meeting the conditions of Theorem 5.1. Find a sentence Z of LS such that $\mathsf{Bew}_{\mathcal{T}}\ulcorner Z \urcorner \supset Z$ is not in $\mathcal{T}$. This means we cannot add $\Box Z \supset Z$ to the list of modal axioms in Section 4.

6 Löb's Theorem

Using the Gödel Fixed Point Theorem, for a sufficiently strong formal theory $\mathcal{T}$ we can produce a sentence that asserts its own unprovability in $\mathcal{T}$ and, as we have seen, it will not be provable in $\mathcal{T}$. The demonstration is basically a diagonal argument, similar to that of Cantor's Theorem, Russell's paradox, or the Liar paradox. The logician Leon Henkin raised the following interesting question. We could just as easily produce a sentence that asserts its *provability* in $\mathcal{T}$; what is its status? A diagonal-style argument won't work now. Löb answered Henkin's question with a remarkable result that says such a sentence will, in fact, be provable. Indeed, we don't

Chapter 10. Gödel's Second Theorem

even need the full strength of having a fixed point, $\text{Bew}_{\mathcal{T}}\ulcorner X \urcorner \equiv X$, in $\mathcal{T}$; it is enough to just have $\text{Bew}_{\mathcal{T}}\ulcorner X \urcorner \supset X$.

Theorem 6.1 (Löb's Theorem). *Again let $\mathcal{T}$ be a Δ_0 complete formal theory, containing true instances of Σ formulas, having the fixpoint property, and satisfying the Löb provability conditions. Let X be a sentence in the language* LS *such that $\text{Bew}_{\mathcal{T}}\ulcorner X \urcorner \supset X$ is in $\mathcal{T}$. Then both $\text{Bew}_{\mathcal{T}}\ulcorner X \urcorner$ and X are in $\mathcal{T}$.*

Proof Assume $\text{Bew}_{\mathcal{T}}\ulcorner X \urcorner \supset X \in \mathcal{T}$. Such a sentence X exists by the fixpoint property, but now we use that property a second time. There must also exist a fixed point for $\varphi(v_0) = \text{Bew}_{\mathcal{T}}(v_0) \supset X$. Let us call a fixed point for this F, and so $(\text{Bew}_{\mathcal{T}}(\ulcorner F \urcorner) \supset X) \equiv F$ is in $\mathcal{T}$. We will use this formula to show that $\text{Bew}_{\mathcal{T}}\ulcorner X \urcorner$ and X are in $\mathcal{T}$.

Once again the argument will be clearer if we use modal notation. Stated in these terms, we have the general modal conditions given in Section 4, and specific formulas X and F for which we have $\Box X \supset X$ and $(\Box F \supset X) \equiv F$. The goal is to derive X and $\Box X$. Here is the argument.

1. Take $\Box F$ as an assumption.

2. $F \supset (\Box F \supset X)$ (because we have $(\Box F \supset X) \equiv F$)

3. $\Box F \supset \Box(\Box F \supset X)$ (regularity rule on 2)

4. $\Box(\Box F \supset X)$ (*modus ponens* on 1 and 3)

5. $\Box(\Box F \supset X) \supset (\Box\Box F \supset \Box X)$ (**K** axiom scheme)

6. $(\Box\Box F \supset \Box X)$ (*modus ponens* on 4 and 5)

7. $\Box F \supset \Box\Box F$ (**4** axiom scheme)

8. $\Box F \supset \Box X$ (from 6 and 7 by classical logic)

9. $\Box X \supset X$ (given)

10. $\Box F \supset X$ (from 8 and 9 by classical logic)

11. X (*modus ponens* on 1 and 10)

Since we have managed to derive X from $\Box F$ as an assumption, using the deduction theorem we conclude that $\Box F \supset X$ is provable. But by our initial conditions, we also have $(\Box F \supset X) \equiv F$, so we conclude F. Since F is provable, so is $\Box F$ by the Necessitation rule. But now we have both $\Box F$ and $\Box F \supset X$, so we have X. And then we have $\Box X$ by the Necessitation Rule again. ∎

Löb's result even provides us with an alternative proof of the Gödel Second Incompleteness Theorem. Very simply, it goes as follows. Suppose $\mathcal{T}$ meets the conditions of Theorem 5.1. And suppose $\neg \mathsf{Bew}_{\mathcal{T}}(\ulcorner \bot \urcorner) \in \mathcal{T}$. The formula $\neg \mathsf{Bew}_{\mathcal{T}}(\ulcorner \bot \urcorner)$ is logically equivalent to $\mathsf{Bew}_{\mathcal{T}}(\ulcorner \bot \urcorner) \supset \bot$, so this must be in $\mathcal{T}$. But then, by Löb's Theorem, $\bot \in \mathcal{T}$, so $\mathcal{T}$ is inconsistent.

Exercises

Exercise 6.1. In the proof of Theorem 6.1 our verification of $\Box F \supset X$ used the deduction theorem. This is a classical theorem, and you probably have not seen a proof of it for the logic K4. Show $\Box F \supset X$ without using the deduction theorem.

7 Gödel-Löb logic, GL

We have been using modal notation as a convenient abbreviation. It is time to treat it more seriously. This leads to the subject of *provability logic*, which is a topic in itself.

Suppose we set up a propositional language in which formulas are built up from propositional variables, P, Q, ... using the usual propositional connectives and an additional formation rule: if X is a formula so is $\Box X$. For instance, $\Box(P \supset Q) \supset (\Box P \supset \Box Q)$ is a formula in this language.

Next, let $\mathcal{T}$ be a Δ_0 complete formal theory in the language LS. We can interpret modal formulas into this theory, as follows. Let v map propositional variables to sentences of LS—call v a *modal valuation*. Extend v to more complex formulas as follows.

1. $v(\neg X) = \neg v(X)$

2. $v(X \supset Y) = (v(X) \supset v(Y))$ (and similarly for other binary connectives)

3. $v(\Box X) = \mathsf{Bew}_{\mathcal{T}}(\ulcorner v(X) \urcorner)$

Let us say a modal formula X is $\mathcal{T}$-valid if $v(X) \in \mathcal{T}$ for every modal valuation v. Then, if $\mathcal{T}$ satisfies the Löb provability conditions (Definition 3.1) the following modal axioms will be $\mathcal{T}$-valid (along with all their substitution instances):

K Axiom Scheme $\Box(X \supset Y) \supset (\Box X \supset \Box Y)$

4 Axiom Scheme $\Box X \supset \Box\Box X$

Further, if all true instances of Σ formulas are in $\mathcal{T}$, then the following rule will preserve $\mathcal{T}$-validity:

Chapter 10. Gödel's Second Theorem

Necessitation Rule $\dfrac{X}{\Box X}$

The modal logic whose axioms are classical tautologies and all instances of the two axiom schemes above, and with *modus ponens* and the Necessitation Rule as rules of inference, is the logic K4. It is quite a common modal logic. An even more common modal logic results by adding the following to K4:

T Axiom Scheme $\Box X \supset X$

This yields the modal logic S4. However, you showed in Exercise 5.1 that this would *not* be $\mathcal{T}$-valid provided $\mathcal{T}$ was strong enough to prove the Gödel Fixed Point Theorem. On the other hand, we saw in Section 6 that under the same circumstances, $\mathcal{T}$-validity would be preserved if we added the following rule of inference.

Löb Rule $\dfrac{\Box X \supset X}{X}$

The modal logic GL (for Gödel-Löb) is the modal logic that extends K4 by the addition of the Löb Rule as another rule of inference. Every formula provable in GL will be $\mathcal{T}$-valid for every Δ_0 complete formal theory that satisfies the Löb Provability Conditions, in which all true instances of Σ formulas are provable, and which has the fixpoint property.

This way of formulating GL is not common, partly because logicians are much happier with axioms instead of rules. If we try to 'internalize' the Löb Rule, we get the following formula:

Löb Axiom Scheme $\Box(\Box X \supset X) \supset \Box X$

It turns out that adding the Löb Axiom Scheme to K4 instead of the Löb Rule also yields the logic GL. We leave it to you to prove this, in exercises.

In the presentation here, the logic GL was connected with provability for theories in the language LS of set theory. As we saw in Chapter 1, Section 5 and subsequently, the hereditarily finite sets can be Gödel numbered, and instead of working with sets we can work with their Gödel numbers—set theoretic assertions become arithmetic assertions. Mathematically, nothing is lost if we confine ourselves to theories of arithmetic instead of theories of the hereditarily finite sets, and the gain is a simpler underlying structure. All theoretical work in the area has been done in the context of arithmetic. It was in this setting that GL was originally introduced, and was subsequently studied in great detail. To do things this way, modal valuations are thought of as maps from modal formulas to sentences in the language LA, for arithmetic, instead of to sentences of LS. Then in the key item

$$v(\Box X) = \mathsf{Bew}_\mathcal{T}(\ulcorner v(X) \urcorner)$$

the formula $\mathsf{Bew}_{\mathcal{T}}(v_0)$ is a provability formula for a theory $\mathcal{T}$ of arithmetic, and not of set theory, and $\ulcorner Z \urcorner$ is a numeral that designates the Gödel number of Z, where a numeral is one of $\mathbf{0}, \mathbb{S}(\mathbf{0}), \mathbb{S}(\mathbb{S}(\mathbf{0})), \ldots$. If we take $\mathcal{T}$ to be Peano Arithmetic, it is still the case that all formulas of GL are $\mathcal{T}$-valid. In a famous (and difficult) theorem, Robert Solovay showed GL was *complete* for Peano Arithmetic: if $\mathcal{T}$ is Peano Arithmetic, a modal formula X is $\mathcal{T}$-valid *if and only if* it is a theorem of $\mathcal{T}$. Loosely speaking, GL captures, in an abstract way, exactly the reasoning about provability that can be carried out in Peano Arithmetic. This is a central result in what has come to be an area in itself, *provability logics*. It is an area that is still growing.

Exercises

Exercise 7.1. (Easier) Consider the modal logic K4 extended with the Löb Axiom Scheme. Suppose the formula $\Box X \supset X$ is provable in this logic. Show X is also provable. Hence the Löb Rule is admissible in this logic.

Exercise 7.2. (Harder) Consider the modal logic K4 with the Löb Rule added. Show that each instance of the Löb Axiom Scheme is provable. Hint: if Z is an instance of the Löb Axiom Scheme, show $\Box Z \supset Z$ is provable in K4, and hence Z is provable by the Löb Rule.

FURTHER READING

The work that has been presented here is only a fraction of what has been done in this area of the foundations of mathematics. The subject is still developing with current work on models of arithmetic, weak axiom systems for arithmetic, and so on. What follows is not about such research—there is much and the literature grows steadily. Instead, here are some very basic suggestions for further reading involving the history of what was presented here, and some analysis of the work in greater detail than we have managed.

Gödel on Incompleteness: Kurt Gödel's original paper on incompleteness is surprisingly readable—surprisingly, since papers that initiate fields can sometimes be confused and tentative. It was published in German in 1931 and good English versions, with commentary, can be found in several places. The best are these.

1. *From Frege to Gödel*, edited by Jean van Heijenoort, and originally published by Harvard University Press in 1967. Currently available in a paperback version.

2. *The Undecidable*, edited by Martin Davis, originally published by Haven Press Books in 1965. This too is currently available in paperback, from Dover Publications.

3. *Kurt Gödel Collected Works*. This remarkable five volume set of Gödel's published and unpublished writings contains the incompleteness papers (in English and in German) in Volume I. Oxford University Press, 1986.

Gödel on GL: The idea of using modal logic to study provability in arithmetic was introduced by Kurt Gödel in 1933 in a short paper, *Eine Interpretation des intuitionistischen Aussagenkalküls* or, in English, *An interpretation of the intuitionistic propositional calculus*. This too can be found in Volume I of *Kurt Gödel Collected Works*.

Turing on Computability: Turing machines were introduced in Turing's fundamental 1936 paper, *On computable numbers, with an application to the Entscheidungsproblem*, and this is a very readable paper. It can be found in several sources. Among these are the following.

1. *The Undecidable*—see item 2 under *Gödel on Incompleteness* above.

2. On-line, at http://web.comlab.ox.ac.uk/oucl/research/areas/ieg/e-library/sources/tp2-ie.pdf

Boolos on GL: A very readable and thorough presentation of work on provability logic, GL in particular, can be found in *The Logic of Provability*, by George Boolos, Cambridge University Press, 1993 (paperback 1995). This replaces an earlier book, *The Unprovability of Consistency* by the same author, Cambridge University Press, 1979, which is also well worth looking at.

Smullyan on Self-Reference: Raymond Smullyan has written many books, some popular, some technical. The following are directly relevant here.

1. *Gödel's Incompleteness Theorems*, Oxford Univ. Press, 1992. An abstract treatment of the subject that is along the lines followed here, but which goes into greater depth.

2. *Recursion Theory for Metamathematics*, Oxford Univ. Press, 1993. A companion volume to the one above.

3. *Diagonalization and Self-Reference*, Oxford Univ. Press, 1994. A detailed study of diagonal arguments, Gödel style, in combinatory logic, and elsewhere, from an abstract point of view.

4. *Forever Undecided*, Alfred A. Knopf, 1987. This entertaining book actually leads the reader from elementary logic to a proper treatment of incompleteness results. And it does it all through a series of entertaining puzzles.

Modal Logic: The present book is not a work on modal logic, but it does come up naturally when discussing the Second Incompleteness Theorem. If you want to know more, a standard treatment is *A New Introduction to Modal Logic*, by G. E. Hughes and M. J. Cresswell, Routledge, 1996.

Handbooks: There are several handbooks with useful articles covering various topics we discussed.

1. *Handbook of Mathematical Logic*, Jon Barwise editor, North Holland, 1982. The chapter *The incompleteness theorems* by Craig Smorynski is directly related to the material covered in this book.

2. *Handbook of Modal Logic*, Patrick Blackburn, Johan van Benthem, and Frank Wolter editors, Elsevier, 2006. Chapter 16,

Modal logic in mathematics, by Sergei Artemov, discusses GL, among other things.

3. *Handbook of Philosophical Logic*, Dov Gabbay and Franz Guenthner editors. The first edition, from Kluwer, consisted of four volumes, 1983-1989. The second and much larger edition, with Springer, began appearing in 2001. In Volume 11 (2004) is *Modal Logic and Self-reference* by Craig Smorýnski, and *Diagonalisation in Logic and Mathematics* by Dale Jacquette. In volume 13 (2005) is *Provability Logic* by Sergei Artemov and Lev Beklemishev.

INDEX

4 scheme, 127, 132

assignment, 81
atomic formula, *see* formula, atomic
axiom scheme
    first-order, 85
    propositional, 84
axioms, 88

bound occurrence, 13
bounded quantifiers, *see* quantifiers, bounded

canonical model, *see* model, canonical
Chinese Remainder Theorem, 43
Church's Theorem, 113
Church-Turing Thesis, 67
closed, 13
completely definable, 124
completely inductive, *see* inductive, completely
computable, 68
concatenation, 51
constant symbol, 12
Craig's Theorem, 90

Δ_0 complete, 89
Δ_0 formula, *see* formula, Δ_0
designates, 15
deterministic, 77
domain, 3, 14
downward saturated

Δ_0, 71
Σ, 72

enumerable, 116
enumerates, 116
$\in$-ordered, 35
equality conditions, 86

first order language, 11
Fixed Point Theorem
    Gödel's, 124
    Semantic, 61
fixpoint property, 123
formal theory, *see* theory, formal
formation sequence, 52, 53
formula, 12
    Δ_0, 24, 40
    Σ, 25, 40
    Σ_1, 31
    atomic, 12
free occurrence, 13
function, 3
function symbol, 12
function-like, 124

GL, 132, 133
Gödel number, 9, 116
Gödel's β function, 42, 43
Gödel's Completeness Theorem, 85
Gödel's Incompleteness Theorem, 97, 105, 109
Gödel's Second Incompleteness Theorem, 129

Gödel-Löb logic, 132
graph, 67

hereditarily finite set, 5

incomplete, 104
inconsistent, 102, 104
index, 74
Induction Principle, 1
induction schema, 92
inductive, 1
    completely, 2
intended model, *see* model, intended
interpretation, 14

K scheme, 127, 132
K4, 127, 133
Kleene's Normal Form Theorem, 74

language, *see* first order language
    of arithmetic, 15
    of set theory, 17
Löb provability conditions, 126, 127
Löb rule, 133
Löb scheme, 133
Löb's Theorem, 131
logical consequence, 83

modal valuation, 132
model, 14
    canonical, 15
    intended, 16
    normal, 86, 87
    standard, 16
modus ponens, 85

names, 15
necessitation rule, 127, 133
non-deterministic, 77
Normal Form Theorem, 31
Notation Convention, 14, 25, 29, 36, 50–52, 56, 88

numbers, 1, 8

ω-consistent, 108
ω-inconsistent, 108
ordered pair, 2
ordinary, 57, 58, 62, 102
    $\mathcal{T}$-$ordinary_G$, 103
    $\mathcal{T}$-$ordinary_C$, 111
ordinary
    Σ, 70

parameter, 84
partial recursive, 68
Peano arithmetic, 91
Post's Theorem, 73
proof, 85
propositional connectives, 11
provability logic, 132, 134

quantifiers, 11
    bounded, 24, 40

range, 4
rank, 6
recursive, 68
recursively enumerable, 68
reflexivity condition, 86
regularity rule, 128
relation, 3
relation symbol, 11
representable, 20
represents, 20
    in a theory, 100
Robinson's Q, 92
Rosser's Theorem, 114, 118
rule of inference, 84
Russell's paradox, 57

Semantic Fixed Point Theorem, *see* Fixed Point Theorem, Semantic
semi-computable, 68

semi-decision procedure, 67
sentence, 13
sequence, 4
S4, 133
Σ formula, *see* formula, Σ
Σ_1 formula, *see* formula, Σ_1
standard model, *see* model, standard
strongly defines, 116
structure, 14
substitution, 13, 54
substitutivity condition, 86
symbol, 50

T scheme, 133
Tarski's Theorem, 59, 62
term, 12
theory, 87
    formal, 90
    true, 88
    with equality, 88
transitive, 33
transitive closure, 37
Trichotomy Principle, 4
true in a model, 18, 19, 82
Turing machine, 75

universal generalization, 85

valuation, 81
variables, 11

Zermelo-Fraenkel, 92

LIST OF NOTATION

$\mathcal{A}(x,y)$, 17
$\mathcal{A}_\Sigma$, 94

BITAND , 9, 37
BITOR , 9, 37

Δ, 68
Δ_0, 24
DIV , 9, 37

FIN_SET, 92

$\mathcal{G}$, 9
GL, 132, 133

$\mathcal{H}$, 8
$\mathbb{HF}$, 17

K4, 127, 133

$L(\mathbf{R}, \mathbf{F}, \mathbf{C})$, 12
LA, 15
LS, 17
L^{par}, 84

MOD , 9, 37

$\mathbb{N}$, 16

ω, 4
$\langle a, b \rangle$, 2

PA, 92
φ_S, 58

Π, 68

Q, 92
$\ulcorner s \urcorner$, 59

R_n, 5

SET_Q, 93
SET_Q_0, 114
S4, 133
Σ, 25
Σ_1, 31
$S \models \varphi$, 83
$\mathcal{S}_\Sigma$, 94
$y\begin{bmatrix}x\\t\end{bmatrix}$, 13
x^+, 4

$Theory(\mathcal{A})$, 87
$t^{\mathcal{M},\mathcal{A}} \in \mathcal{D}$, 82

$x^{\mathcal{A}}$, 82

ZF $- \infty$, 92

DEFINED FORMULAS

S-1 (x subset y), 25
S-2 (x equals y), 26
S-3 (x is $\varnothing$), 26
S-4 x is $\mathbb{A}(y, z)$, 26
S-5 (x is $\{y, z\}$), 29
S-6 (x is $\{y\}$), 29
S-7 (x is $\langle y, z \rangle$), 29
S-8 Ordpair(x), 29
S-9 (x is $\langle y, \_ \rangle$), 29
S-10 (x is $\langle \_, y \rangle$), 29
S-11 Relation(x), 29
S-12 Function(x), 30
S-13 (x is Domain y), 30
S-13 (x is Range y), 30
S-14 Transitive(x), 34
S-15 Number(x), 34
S-16 (y is x^+), 35
S-17 (Sequence x With Domain y), 35
S-18 Sequence(x), 35
S-19 (z is x_y), 35
S-20 (z is $x + y$), 36
S-21 (z is $x \times y$), 36
S-22 (z is $x \uparrow y$), 36
S-23 Variable(x), 50
S-24 (h is $f * g$), 51
S-25 TermSequence(s), 52
S-26 Term(t), 52
S-27 ClosedTerm(t), 52
S-28 Atomic(s), 53
S-29 FormulaSequence(s), 53
S-30 Formula(f), 53

S-31 (Formula f With Free s), 54
S-32 (u is $s\begin{bmatrix}v\\t\end{bmatrix}$), 54
S-33 (B is $A\begin{bmatrix}v\\t\end{bmatrix}$), 55
S-34 (Y is $X\begin{bmatrix}v\\t\end{bmatrix}$), 55
S-35 Names(t, s), 56
S-36 RepresentingFormula(f), 56
S-37 (X is $Y(t)$), 57

Δ_0-DownSat(v_0), 72
Δ_0-True(v_0), 72